YAKUB

(Jacob)

The Father of Mankind

By

Elijah Muhammad

(Messenger of Allah)

Compiled & Edited by

Nasir Makr Hakim

Minister of Elijah Muhammad

Published by
SECRETARIUS MEMPS PUBLICATIONS
111 E Dunlap Ave, Ste 1-217
Phoenix, Arizona 85020-7802
Phone & Fax 602 466-7347
Email: secmemps@gmail.com
Web: http://www.memps.com

YAKUB: THE FATHER OF MANKIND

Secretarius - ii

WWW.MEMPS.COM

First Edition

EAN13# 979-8-4004-0002-5

Printed in the United States of America

DEDICATION

TO ALL THE STEADFAST
BELIEVERS AND FOLLOWERS
OF
ELIJAH MUHAMMAD,
MESSENGER OF ALLAH,

~ ~ ~ ~ ~

THE
MESSENGER
IS A SWORD
IN A MAN'S HAND

~ Sayings of Elijah Muhammad, Vol. 3

TABLE OF CONTENTS

PREFACE

I have been publishing Elijah Muhammad works for the past twenty years. My initial intent was to publish for children, especially since I had acquired my second degree in Early Childhood Education in California. I did not see many books addressing the special needs of black children. After ascertaining a concrete cause as to why this was so, I began purchasing the necessary computers to develop materials to satisfy this vacuum. Another task I began to undertake was collecting as much documentation of Elijah Muhammad that I could find, and please believe me, at that time, the prices were high. Yet, like the Saviour, Master Fard Muhammad's father, I got what I needed regardless of the price. The price didn't mean a thing, because I wanted to assist in the restoration process of my people, and no price is too high. But little did I realize at this time, that I would eventually pay with all I had.

As I set out to develop the idea of publishing for children, I had amassed more and more of what Elijah Muhammad taught. The materials collected ranged from audio and video tape, articles from early writings and the Muhammad Speaks newspaper. Occasionally, I would be blessed with more intimate writings from the Messenger's own hand as well as letters from persons whom the Messenger had written. Although these documents and media are the base of what I have collected, more important, is my belief in Allah, Master Fard Muhammad, and His Messenger, Elijah Muhammad, because without belief in Them and the power of their spirits, this collection would be like having the best transportation on the planet without the keys to use it.

My initial submission to my first mentor and self-experience through traveling nationally around those of the Nation of Islam, laid the basis for a seed bed, from which to mature in the knowledge I had gained. My interaction with pioneers in the Nation of Islam, who had the pleasure of the Messenger of Allah's presence, added an unparalleled "seasoning" to what was once a shallow breath. With what I have learned from an abundance of materials and selected interaction, has virtually cut years off of my learning experience; thus, relieving me of needing to reinvent the wheel. These experiences are in and of themselves priceless.

After developing an appreciation of Elijah Muhammad's thinking, this knowledge convinced me to change my focus from publishing for Allah and His Messenger; for I had initially set out to do my will, but the more I learned, the more I began understanding my duty, which took precedence my to His Will (All Praise Is Due To Allah).

One of the many experiences I would like to share with you, the reader, is an instance when we would read over the Messenger's teachings numerous times and would get different bits of knowledge that we somehow "missed" a previous time. Many of us can relate to this experience; yet, there are so many others who feel reading a book once, not to mention twice, as drudgery. These are the ones who would write the Messenger complaining that he didn't put "new" articles in the Muhammad Speaks newspaper on a timely basis. The Messenger would respond to the effect that their actions had not changed to signal that they had caught up with the first ones. I have read letters that the Messenger's books seem to have the same materials in them; therefore, buying more is unnecessary.

There are various lessons in different schools of thought that teaches against giving a novice certain knowledge before time. The Holy Qur'an admonishes us to not ask questions until a certain time. What I am getting at is this: Without a certain level or breath of knowledge, we may treat the revelation of knowledge disrespectfully, because he have not developed in knowledge. If we haven't developed a background with which to receive new knowledge, we may treat it carelessly out of ignorance. This has been the case with Elijah Muhammad's teaching. Many have not taken it seriously enough. This is exhibited in the families and lifestyle by many who have amassed his teachings, but they don't know how to apply them. I could have said this more euphemistically, but if they have them and yet will tolerated the awful conditions they are experiencing, then either they're simply ignorant and don't know that they don't know or they are fools who know that they don't know, and proud of it.

This book, Yakub: Father of Mankind, is not the only attempt at this subject; however, I wanted to take the nine sources of the Messenger's own words and compile one book dedicated to this subject; a subject which Wallace Muhammad, Elijah Muhammad's son had banned from being taught when he first took control of the Nation of Islam. This subject was also butchered into a fairy tale by Minister Farrakhan and eventually discontinued.

Elijah Muhammad writes, "...You are bearing witness to the people that I am the man from God. You are to show proof that compares with predictions. You have not seen God, but you believe that I have seen Him....We are building a new world. I have to follow close to my Lord [Master Fard Muhammad]. You have to follow close to me to follow Him...You cannot take Yakub's history and defend it as I can. It is best that you teach the people what you heard me teach."[1]

To this end, I ask your patience with this book. It may appear that the Messenger repeats himself in certain instances; however, if you are not careful, you can miss very important bits of information necessary to see into how the Messenger can take current phenomena to substantiate the creditability of Yakub's existence and history, especially in a time and amongst a people who see its revelation as most dangerous.

I am aware that some will say, "I've read this before, but how many pieces of the puzzle were you looking at all at once?

The profundity of the Messenger's works command a certain approach when putting it before the people. It

[1] Muhammad, Elijah; Letter to the Minister, March 11, 1965.

deserves to be in a form consistent with the high caliber of its content. This is why after seeing a shallow attempt at putting a document out about Master Fard Muhammad, I felt compelled to challenge such a mediocre approach with The True History of Master Fard Muhammad. This attempt was to first of all address the various attempts by non-believers and hypocrites writing about that which they either didn't believe or was unqualified.

I secondly pointed out the fact that they all clearly stated in their books a consistent mind-set of uncertainty about Master Fard Muhammad's whereabouts or origin, disqualifies them all from being taken serious about the subject. I lastly pointed out that the only ones who benefited most from Master Fard Muhammad's presence was Elijah Muhammad himself and the believers in the Nation of Islam. This alone unequivocally makes them the best qualified to speak on the matter. After additionally, challenging, dispelling and completely refuting the notion that Master Fard Muhammad and Elijah Muhammad somehow borrowed or was taught by Nobel Drew Ali and Marcus Garvey, The True History of Master Fard Muhammad, gave, according to national feedback, the best representation of what Elijah Muhammad taught, other than he himself. The basis of this book will pursue the same course; that is, first address the present notions surrounding Yakub and then give you what Elijah Muhammad teaches of him.

The above is crystal clear; however, what is it about this history that made Wallace Muhammad, Elijah

Muhammad's son, once Registered Muslim, order the Ministers of Elijah Muhammad to stop teaching it in 1975. Even the current regime under the organization of Sheik Farrakhan, in Chicago, doesn't teach it now that he has become like Wallace, "recanters" of Master Fard Muhammad as God in person and Elijah Muhammad, as His Last Messenger.

In the news are different references to animal and human cloning. What we must understand is that years before they told you, they have been secretly working on this process. This tells us many things: 1. They got the idea from somewhere. 2. What does the deformed experiments look like and what happens to them when they are disposed of? 3. How long has this been going on? 4. Who funds this project? The list goes on and on. Yet, just as this is accepted as the people's reality today, there was a time when it would have been considered far fetched or unimaginable. What I am suggesting is that, just because an average working or unaware person, whose day to day routine is predictably mapped out, doesn't know that these type of projects exists, it doesn't mean they don't. Just because a person with an ordinary, controlled life doesn't look up or doesn't see these hidden truths on CNN, ABC, NBC or Nightline, doesn't mean that it wasn't happening before they were born. On the contrary, according to Elijah Muhammad, it was how the white man and woman was made. This is why they are called "mankind." They are not original. They are a "kind" of an original made from an Original man, in the Original man's image, after the Original man's likeness.

When asked how does one substantiates or supports the validity of an occurrence, phenomena or event that happened thousands of years ago? How does one prove the existence of an incident, from which, the chief products are people who want to hide and deny their origin? How do you confirm the existence of a project, from which a world is produced that extremely oppose exposing its origins? How does one expose the identity of a man, the knowledge of whom, was hidden for 6,000 years?

The answers to these questions must be answered in the context of the person who originally revealed this knowledge. It originally came from God, Himself and was taught to Elijah. This is very consistent with what Jesus said when he stated, that to know that truth, it would set you free, but little did we know that the world in which we lived is manufactured based on a lie. This would explain the awesome power of truth and adverse affects it would have on this world.

To answer these questions would mean proving that Elijah Muhammad is who he says he is, which would mean proving he is consistent with the prophets. The prophet's consistency is the basis of their creditability and their creditability is consistent with the natural laws governing the known universe, and it's consistency is what serves as the basis of our belief that there is a higher power. In other words, each expression of creditability support, upholds and substantiates the creditability of another. Therefore, in order to substantiate the creditability of this book, it must come from what Elijah Muhammad said. His creditability will verify it is from God Himself. Of all those who think they know, He is the Best Knower.

As we compile what the Messenger teaches about Yakub from different sources, we will footnote key aspects to prove it modern usage, not necessarily the validity of the truth. The truth is sufficient to prove itself. Once we eliminate what it's not, then all that will remain will be the truth or the answer. This is what makes Elijah Muhammad teaching superior; hence, the Supreme Wisdom.

I am confident that the reader will benefit greatly from this approach. It will help them figure out many things for themselves and in this way, they will begin to own or master it; for Elijah Muhammad stated that it is useless to give you something of wisdom if you can't master it. From the look of the current state of affairs, few are mastering this great wisdom. Knowledge is the result of learning and is a force or energy that makes its bearer overcome obstacles, barriers or resistance. In fact, god means possessor of power and force, so teaches the Messenger. It's time to "learn," now that we have gotten all we could amass of the wisdom. Let's stop going "by" the books and go into the books and use it to secure the promise of our Lord, while following closely to him following his Lord.

Lastly, don't let the appearance of redundancy fool you into thinking that you understand that which you know not. Although we see night and day overlap themselves all of our lives, we still miss so many lessons, because we have had the wrong teachers who didn't teach us how to see or hear. Like Jesus said, "Having eyes but cannot see. And having ears but cannot hear." The Allah says in the Holy Qur'an, that, "In the alternation of the night and the day, there are signs for those who understand." The night overtakes the day and the day over takes the night, but do men mind?" Don't be blinded by what appears to be redundancy!

Your Brother
Nasir Makr Hakim
Minister of Elijah Muhammad, Messenger of Allah

In The Real Beginning . . .

Of all the knowledge that one may have or obtain, the knowledge of God is the greatest and the most necessary of all knowledge. I would say that 98% of the people of the earth are without knowledge of the God, the Supreme Being.

The Christians and most of the old world Muslims are alike; not having a true knowledge of the Supreme One, referred to as Allah, and God makes most people believe that God is something other than a man.

Looking at God's creation (the universe) and His creatures without number and unlimited, we have never been able to obtain the knowledge of just how God created the Universe and Himself.

We have never known where He Himself began or who was first, God or the Universe? Just who is God? What is He? Is He of the essence of this universe or some invisible power, spirit, or force that has no equal comparison in His creation?

The wisest of scientists have worked, studied, and searched all their life long for the actual knowledge of God and have failed to obtain it. Finally, they formed their own opinions of God.

Many claim that there is no such one (God) and that the Creation of the Universe came by chance. Man's creation, they say, took place from some lower form of animal life of the sea. But they are never able to prove such theories. Therefore, the people formed their own ideas of God, for they had no true knowledge of God.

How can we know God if He kept Himself from our having a true knowledge of Him? Just what was the purpose of this secret knowledge of God?

We just could not know God unless God made Himself known to us, and there was a set time for Him to make Himself known. Why was there a set time?

There are not any gods Who live forever. Their wisdom and work may live six thousand or twenty-five thousand years, but the actual individual may have died within a hundred or two hundred years, or the longest that we have a record of, around a thousand years.

There is no God Living Who was here in the Creation of the Universe, but They produce Gods from Them and Their Wisdom lives in us.

The knowledge of God has been kept a secret by twelve men on our planet for many thousands of years. The twelve pass their knowledge on from son to son, but the number possessing this knowledge is never more than twelve; and they are not to ever reveal it. (OSHA 61-62)

Human Beings are created according to the life of the Universe (planets). Master Fard Muhammad, to Whom praises are due forever, taught me that the Original Designer and Maker of the Universe created it on time and there is an end to it.

Scientists have learned by study that everything we see that we call Universe is not endurable. It is gradually decaying. The Bible and the Holy Qur'an both verify this decay of the Universe and that one day a Wiser God than Them all will exist in a new Universe. A new universe means that it will not be exactly like this one that we know. According to the hints of the Holy Qur'an, it will be a better one than we have. Naturally, after experimenting with a thing that has been made, we can improve on a new make of it. (OSHA 96-99)

The Wisdom of Each God, according to what Allah (God), in the Person of Master Fard Muhammad, to Whom praises are due forever, taught me, has a cycle of twenty-five thousand years. Once every twenty-five thousand years, another God would be given a chance to show forth His Wisdom to the people.

This has been going on for many trillions of years, according to His teachings to me -- at least since the deportation of moon and earth, according to the science of why we use a cycle of twenty-five thousand years, this is compared to the circumference of our earth at the equator, which is approximately twenty-five thousand miles. This is also compared with the rotation of our earth or the change at the poles which also makes a complete change once every twenty-five thousand years. (OSHA 96-99)

Let's now consider this a little deeper. In the first place, one God was the God of the Universe. In the last place, One God is the God of the Universe. Everything that is of life and everything that is of metal or everything that is matter in the Universe came from Him. He created it! (OSHA PG42)

He [Himself] is Self-created. How could Self create Self? Take your magnifying glass and start looking at these little atoms out here in front of you. You see that they are egg-shaped and they are oblong. You can crack them open and you will find everything in them that you find out here. Then were there some of them (atoms) out here? Well who created them? I want you to accept the Black God. You say, "There is no beginning or ending." I admit that. But we do know that they had to have some kind of beginning. But how it happened, we don't know. That's why we say that His Beginning, we don't know anything about.

Imagine, you close your eyes now; imagine the whole entire visibility of the Universe is gone from you. This is the way He was born; in total darkness. There was no light anywhere. Out of the total orbit of the Universe of darkness there sparkled an atom of life. Once upon a time; but don't ask me when it was. How could that atom of life make a record of it's own creation? It could not write it's own creation, the record of it, because He was the First; there was no recorders around Him. He was First to record His

Ownself. How long was that? We can't tell; we weren't there. He was the only One in the whole entire dark Universe. He had to wait until the atom of life produced brains to think what He needed. How long was that? I don't know, Brothers. But He was a Black man! Coming out of total darkness at that time, we all could say that we are produced by a white god, but there was no light nor any white anywhere; there was All Darkness. So God revealed to me. In that Darkness, which had no end to it there - that Darkness Created an atom of Life, and the Color to be the Color had to be Black, as there was no light; therefore, it had to be the Color of the thing that Created it! All Praise is due to Allah! (OSHA PG39-40)

Never has this kind of teaching been poured into your ears since the Earth was or the Nation was. I teach you not of prophets, but of God? No, no, no. I am not teaching you of Isaiah and Moses and the others. Those are prophets. I am bringing you the knowledge of God Himself!

The Black man's God. Why do you reject the Black God when the Black God is Your God! The Black God made the white god! Why do you reject Him? Long before ever there was - as the old people used to tell me when I was a little boy - a when and a where, He was God. I bear them witness today. Just think it over; a little small atom of life rolling around in darkness. Think it over; Building itself up, just turning in darkness, making its own self. Let's go into it. Do I have any proof of this? Yes, sir! He made Himself into total darkness. He put His Ownself turning, turning on His Own Timetable in the Black womb of the Universe. He

started rotating. He demands every life that comes into the Universe today to start turning first. "Over to Me, for I had that to do myself." "Now I am going to see, that every life that comes into this Universe comes out of total darkness - every life!" Yes sir! It must come out of total darkness. Out of the womb of our mother did we come. We were created there out of the sperm that was emitted into that total dark room, the womb. It took that to make that child. He couldn't be made in the light. He must be made in total darkness.
We must know ourselves, we must know the Nature of Life. We must know the law of Life before ever we can say that we are Masters.

Now, if, by nature, we are born in total darkness, think it over. And, if, by nature, life brings us out of darkness; and, if, by nature, we walk in light from out of total darkness; and, if, by nature, we can think through darkness, and bring light out of that darkness by our own brains, we did so with the white race. Look right into the sperm of life and find him - here's an unlike. Take it! separate it! Your first separation from the white and Black was done. (OSHA PG43-44)

It was a Black God! That is Who did it. He has taken His Universe, (think it over), and engulfed it with the path of light that you call the Milky Way. Think that over. He engulfed His own Universe with nothing but a path of Light. He made the lights so close to each other that the light of this one never stops before this one ties his light in with that one. It was a whole path of nothing but light, a belt around His Universe. He thought so much of His Earth,

He loved His Earth so well, that he took His own Earth and divided the live planets rotating around it with light - stars. Think it over. About 600 - million miles out from this planet, He has a region of stars that divides the other planets from these inner planets. He has made a wreath around His own Head of nothing but stars. Think it over. He made a crown for Himself out of the Earth with stars. "This is My House, and I want to put a crown around it with stars." (OSHA PG44-45).

Look at the ball of your eye. Think it over. That ball of your eye is white, all around a black. A Black sphere there, sitting, surrounded with white. Not the white in the center of this ball; it's Black here in the center. But white around the Black. Right? He has made it like His Universe - around Him is light, but He, Himself is Black! All praises are due to Allah. Look into His Eyes. His eye tells you who He is. And all the offspring from Him bear witness that I came from that Black center there in His eyes. They bear witness that their eye testifies to the truth that it must have been grafted from yours because it is not the equal of your eye. All praises are due to Allah! Would you trade your beautiful Black eye, surrounded with the beautiful white, for a green-blue eye? Would you do that? No, you won't. It is a more beautiful eye, Black. Clear Black, and clear white around it like His Universe He made. He was Black inside, and then He said, "Let there be Light - "On The Outside." It is an unending teaching you about your Black God. (OSHA PG52-53)

Thy Will. In the making of God Himself, He could not have had a Will until he had brains capable of thinking. He was created, self-created from an atom of life [who as well] produced flesh, bones and blood from the earth that He was created on - self-created. I want you to know that good, and I will repeat it time again and I will go to no limited argument with you that He was made on the very earth that we are on now, but it was not exact as it is today. It was only atom itself. (TOT Pg188)

Creator means one who creates a thing, not one that borrows the essence to build something [from] the essence that was made by someone else. You [have] to create your own essence; that's what God did. He did this out of nothing. The whole atmosphere was void of any matter until he made matter in the atmosphere. (TOT Pg. 284)

I say, love your God. He is a Black God and He is forever and forever. There is no such thing as trying to destroy the Black man. You can't do that. It is impossible! One of Our God-Scientists, 66 trillion years ago, as God taught me, tried His best to satisfy His mind by trying to destroy us. But we are here today, and we see His idea right up there - they call it the Moon. That is what He tried to do. Now there is no life on that Moon. He didn't allow anyone to be left to write His days of Himself and His great Wisdom of Himself. We wrote it! We will do the same by you. You won't be able to write our doom, but we can write yours. All is set for this thing - before you ever were created or made. He looked into you before you were born, saw your thoughts, listened to what you would be saying today and saw your works. We just fold up our arms, sit down, and wait for Time. Allah said, "They just waited. They're so glad that the day has come now that they can show you now that the Black man is God!" (OSHA PG 56)

THE SIGN OF THE MOON

The blasting away of MOON by an enemy that robbed the MOON of its life (water) and poured it onto this part of the planet Earth can also be compared to the evil white slave-traders, guided by John Hawkins, the explorer, to come among us and take us by force and bring us to this part of our Earth among strangers whom our fathers knew not. In so doing they made us to lose all knowledge of ourselves and our kind, like this man or God who in his frenzy to try to force all the people to believe as he believed and to speak the same language with no difference in dialect, caused the deportation of the MOON from the Earth. So it is with our enemy today. He wants us all to kneel and bow to his way of life and if he cannot get us to do so he wants to be rid of us and our Earth. (FOI-14)

We see then, that Allah used the history of the work of the God of the MOON, sixty-six (66 trillion years ago, to teach us of something of ourselves and of our enemy. Here, we the Black man in America stand as a MOON devoid of wisdom and the knowledge of self and others; and the MOON is devoid of water. We were robbed by a mastermind man with the purpose in mind to destroying the life of the Black man. That is what he began with. In the history of the making of the white man, Yakub, the maker of the white race, taught his people to destroy the Black babies in the very beginning and this is his ultimate aim (to destroy all the Black people of the Earth). (FOI-23-24)

Do not forget that the intention of the two Gods (the God of the MOON and the God of the white man-Yakub) are the same. The white man would by no means bring us away from our native land then restore us again back there. He did not have that in mind when he brought us here. Nor does he have it in mind to help us; but since trouble has arisen in his own house, he is willing to throw us out rather than take the suicidal step of trying to kill us, which will also cause an immediate destruction of his own life because the God of the Lost-Found members of the Aboriginal people is present and He would destroy such plan and bring to a naught and destroy such planners who would try to destroy us. Nonetheless, it is in the mind of the white slave-master's children to destroy us. (FOI-26)

We, the Black people, are the people out of which Yakub, the father of the white race was born. Yakub was the same as we, until he entertained the idea of making an enemy for us. Yakub himself was not an enemy but his wisdom, knowledge and know-how, to make an enemy for us caused him to be classified then as being a chief devil or enemy of the righteous. (FOI-31)

The Black God produced Himself; He's Self-created. Then that Black god made the white man. He didn't create him. He made him from Himself. If He had made him or given him the power to create himself, He could never rule him. He is doing a hard job at it now. The Bible teaches you in Genesis that God said; "Let Us make man." I want you to wake up. Well if somebody is talking about making a man, why did He use 'Us'? The Holy Qur'an says; "We made man." We! There's no difference in "Us" and "We." Who is that "Us?" Who is this "We?" You say, God made the man. Well, God did make the man. But why did He have that "Us" and "We" around Him? Because that's impossible now. God can't make a man without a help. No, no! He was Self-created Himself, but He can't make a man without help now because the law of Nature, Creation, won't allow Him to do so. So it takes an "Us" now. It takes two people to make a man. Therefore the Bible and the Holy Qur'an plainly teach you, "Us" and "We" made him; did not create him, but made him. (OSHA PG41-42)

You don't say that you created this metal here, because the essence of it is out there in the earth. You didn't create that. But you take it and melt this, the metal, out of something like stone or sand. You can scoop it up down there and sift it - but it was already out there. Then you make whatever you want to from that metal. But, now, go and create something! (OSHA PG42)

Since we started from nothing and since God Himself started from nothing and we're going to something, because the Creator found something more in the darkened circle than he had thought that He would find. He's not here today - that same man - but He laid a base for a wise God to come and get a platform out of His work of self creating.

WHY GOD IS ONE

Once every 25,000 years, (everything I say is figured right), there's a new God that's been coming up in the past, so God taught me. Their wisdom would always run through about 25,000 years and then they would change and bring in another one. All the Gods that rule, as you didn't know, rule by ones and not by just the same one. This is why you are taught God is one - that's right, but it's not the same God. They all [ruled] with righteousness until about 6,000 years ago. They produced a contrary God that wanted to make his name known in power and wisdom in the universe. This was only 6,000 years ago and this was taken up by one God to rule the people for 6,000 years. So, 6,000 years we had a black god, his name was Yakub. (TOT60-1)

These are the Days of setting up a clean world under the guidance of the Original People (Black Nation) who were never guilty of doing evil until one of the gods by the name of Yakub discovered the essence of the Black Man to make a white man and Yakub did just that.

By One Man Sin Entered Into The World

Now since he has discovered this essence, today we are under the God (Allah) to do away with that very essence in us so that no other man can make another people who are different from us. This will never happen again. That is why you are taught that you will be born again physically. This rebirth is in order to get rid of that wicked material in the very essence of the sperm of the Black Man. This wicked material will not be in the Black Man for any one to use after the removal of the present world and people.

You will be caused to grow into a new person and the nature will be different. It will be the nature of righteousness and then we cannot sin. In that kind of rebirth, it will be impossible for us to sin.

The world (of the white race) is doing what it was made to do -- to try the righteous with wickedness and filth. The white race cannot do any better than what they are already doing. So we do not argue and quarrel with that, which they, by nature, are made to do. I repeatedly teach you this -- that you cannot change the nature of the white man unless you graft him back into that which he was grafted out of. This is what Jesus means in the Bible, St. John, 3:7, in his conversation with Nicodemus. He told him that in order to enter the kingdom of heaven he must be born again because in John 8:44 (Bible), Jesus had condemned all of the white race to be devils, and their father was the devil. There was no good nor truth in the father that made the white race. So when a thing is what it is by the nature in which it was made or created in, you do not change it unless you go altogether back to the material that it (he) was made out of. So this demands a rebirth. Jesus was right -- a rebirth -- born again, all over. Some of the clergy and the scholars of Christianity take this to be a spiritual rebirth. But it means that the actual flesh and blood that was discussed in this conversation between Jesus and Nicodemus has to be changed. Then when the man has been put back into what he was taken from, the spirit of that which he was put back into will come to him as the spirit to do evil comes to him now in what he is made in. For he was made out of evil. The white man was not made to obey Allah (God) and to seek after His Righteousness. So, therefore, to make the white man one of the righteous, the white man has to take on a new birth -- the flesh and the blood has to be changed. (OSHA PG 77)

Let us take a look at the devil's creation from the teachings of the Holy Qur-an:

"And when your Lord said to the Angels, I am going to place in the earth one who shall rule, the Angels said: What will Thou place in it such as shall make mischief in it and shed blood, we celebrate Thy praise and extol Thy holiness." (Holy Qur'an Sharrieff 2:30).

This devil race has and still is doing just that - making mischief and shedding blood; and it was the Black Nation from whom they were grafted:
"When your Lord said to the Angels: Surely I am going to create a mortal of the essence of black mud fashioned in shape." (Holy Qur-an Sharrieff 15:28).

The essence of black mud (the Black Nation) mentioned is only symbolic, which actually means the sperm of the Black Nation; and they refuse to recognize the Black Nation as their equal, though they were made from them by a Black Scientist (Yakub). (Foundation II, pg. 133)

We used to say, and many of us still say, that Yakub was the first God of murder, which is true of his race (the white race) that he made. Now, the God who tried to destroy us along with himself, his effort to do so, caused the making of our MOON. The God of the MOON had murder and the destruction of the human family in his mind. He was not successful in doing so. Allah (God) used this history of that God to teach us what is in the mind of this people (the white race). (FOI-9)

Yakub, the father of the white race, did not make the white race from nothing. Yakub made the white race from us (Black Man). So he took living material (Black Man) to bring out a new people. And so Will Allah, the God of Righteousness. Allah (God) Will Take living material, the so-called Negro, and make a new people out of them. The so-call Negro, the Black Man, up from slavery of four hundred (400) years, here in America, is the people out of whom Allah (God) Wants to build a new heaven and a new earth or we say, a new people and a new government. (OSHA PG116)

Before the making of the white race, we never had their type of evil people. The Black Man was never under an evil rule. Evil was never practiced among the Black People before the making of an evil world by Yakub. We never saw or experienced a civilization like the white man's civilization. We never had an unalike people among us before the white race was made. We were all alike especially in color. (OSHA 119-120)

The new man, the white man, came from us, but he is different from us. After Yakub grated his man (white man) from us, his man became a new man to us. We are not a part of the white man.

The white man just looks like a human being, and he is a human being, but he is not kin to us at all. You say, "I cannot understand how that is, Mr. Muhammad." This is true. After the white man was grafted completely out of the germ of the Black Man, the White man was made into a new man, different from the Black Man, from whom he was grafted. The white man became a new person altogether and his very nature is new to us. They do not have the same nature as we have. The white man is different by nature than the Black Man and the white man has made many things new. (OSHA 119-120)

The White Race is an unusual people and their ideology is unusual. He has and was given the gift of a creative mind. To allow him to use his own ideas, the Black Man or Gods were put to sleep in order that the Wisdom of the Black Man did not interfere with what the white man is made for (to rule us under wickedness, enslavement, deceit, murder, and death for six thousand years). [OSHA 96-99]
The Black Man's Gods, according to the history He taught me, have All been the Wisest. They made the white man after their order in wisdom except the knowledge of how to bring into reality and perfection their vision and idea of what they want to perceive equal to the Black Man's Wisdom. This was kept back. They are forced to build their world on the basis of what they found in the Wisdom of Black Man. The white man only made, formed, and put into service those things that met with the carvings and necessities of his people.

The white man was given the power of vision and different ways of life to enable him to build an unalike world from what we have had throughout the millions, billions, and trillions of years.

No God Who is going to rule the people of earth universally, as the white man has ruled for the past six thousand years, was to be given a history or knowledge of the God Who ruled the people before Him. This is in order to keep the Present God from patterning after the Former God and to force Him to use His Own Wisdom in making a world and not a world patterning after previous Wisdom of the Gods Before Him.

The white race did not get a chance to rule solely according to their wisdom, because we were present and they patterned much of their world after what they saw of the Gods of the Black Man. This was due to the fact that the white race was made an enemy to the Black Man; therefore their time to deceive the Black Man was limited to a short span of six thousand years that are used by the Black Gods to rule. (OSHA 96-99)

For the past six thousand (6,000) years there was no god present who was superior to Yakub. Our God and Scientists were not permitted to interfere with the people of Yakub and their civilization or in the way in which they were thinking. That is why they (white race) would kill prophets who came among them teaching the way of righteousness to the people because Yakub's people, the white race, was not to build a civilization on the basis of Righteousness. The white race was to build a civilization just the opposite of Righteousness, and this he has done. (OSHA 123-124)

THE CURSE OF THE WHITE RACE

The secret of the White Race, just couldn't be exposed before their time of rule was out. If the Original Black Nation had known that this new, unalike, blue-eyed race would not be accepted by the God of Righteousness, at the end of six thousand years of their time, there would not have been one left on our Planet. The knowledge of them was kept as a secret, from the common people, as they are symbolically spoken of under the name of "Cain," who murdered his brother. God marked Cain so that the people would not recognize the murdered of their righteous brother, lest the people would kill him." (Genesis 4:15)

Though God pronounced a "Curse on Cain, coming from the Earth, which had opened her mouth to receive the righteous blood from the murderer's hand, he was to be driven from the face of the earth and from the face of God. The earth shall not yield her strength to Cain; and he shall be a fugitive and a vagabond." (Genesis 4:10, 12).

Moses reminded them of this curse in Deuteronomy (28:45-52). Of course, the so-called Negroes are warned in the same Chapter, to let the Divine Chastisement or Destruction of their enemies serve as a lesson to their future greatness, or they will meet with such fate. (Foundation II, pg. 149)

Our father created and made the universe, and one of us made them. As we made the present universe, we will build a new universe. Can the white scientists do the same? We played sleep while they worked; now we are awaking and will build a new world which will make yours look like child's play. We have above you seven inhabited worlds - they are with us These seven worlds or heavens, the stars and the moon show the weight of our brains. What do you have that we did not give you? You shall soon come to know who is the wisest and the most powerful. (Found2, pg. 222)

HOW THE BLACKMAN RULES

We, the original nation of earth, says Allah, the Maker of everything -- sun, moon and stars and the race called white race -- are the writers of the Bible and Qur'an.

We make such history once every 25,000 years. When such history is written, it is done by twenty-four of our scientists. Once acts as Judge or God for the others and twenty-three actually do the work of getting up the future of the nation, and all is put into one book and at intervals where such and such part or portion will come to pass, that people will be given that part of the book through one among that people from one of the Twelve (twelve major scientists) as it is then called a Scripture which actually means script of writing from something original or book.

There is a significance to the number 24 Scientists and the 25,000 years. The number twenty-four Scientists used is in accordance with the hours in our day and the measurement of the circumference of our planet around the equator and in the region of our Poles, Arctic and Antarctic Oceans.

Our planet is not exactly 25,000 miles in circumference, it is 24,896 and we, according to astronomy, don't have a full 24-hour day but near that--23 hours, 56 minutes and 46 seconds. The change made in our planets rotation at the Poles is about one minute a year and takes 25,000 years to bring about a complete change in the region of the Poles.

The actual Poles are inclines 23 2 degrees to the plane of its orbit. The original black nation used 23 scientists to write the future of that nation for the next 25,000 years, and the 24th is the Judge or the one God, Allah. Allah taught me that, once upon a time they made history to last for 35,000 years. (MTTBM Pg108-9)

It was six trillion years, so God taught me, before He put into you the sun, but after the sun then other Gods were born and they thought up the same idea; make something out of fire. Never have you known what and how fire came into space and that stars of fire and a big star we call Sun is nothing but a ball of fire. I want you to remember these things so that you don't make mistake.

If the God created Himself out of matter in which still exists here, then he taken it out of the darkness of space, how then can we declare that the fiery stars - along with our Sun - was created out of space - where did the fire come from?

So today everything is full of fire. I'm full of fire you're full of fire. "This is our figure here?" No, I'm going to tell you here today, but you keep coming back I will tell you, but if I tell you everything you will forget mostly all of it, then you will start adding in your god science and I won't remember what I said after you get through with it (SMILE).

THE SCIENCE OF NUMBERS

Our number is 6, which is from that which we began to make to revolve as He was. He made Himself to revolve and then He caused whatever comes in the darkness to revolve like Himself and that He's doing. We can't help from revolving. We are not a perfect human being, because actually our form is not perfect. So since this is our number here what were we before that number? It was like this, we left out of nothing "0," and we made ourselves into "1." Now if we made our selves, in the beginning, into "1," here comes this nothing "0" again and it goes over and here stands on the right side of "1." Well, if you had 400 years to wait, I think you still have a little time. We were in America for 400 years before the coming of Allah to us.

One "1" comes back and gets behind nothing "0" and makes that nothing there increase it's power 10 fold. In 400 years we were made nothing. That's right, you were everything the white man made you or taught you. After these 400 years as nothing "0," One has gotten behind it, and that One is Almighty God Allah. It made us 10 fold the power and value of that which we were before He came, right? This is because the power of this "1" is generated into nothing "0" and made nothing become something. If I had moved this nothing "0" over here in front (left side) of that "1," it would still keep this as a fraction of nothing to build, but it didn't do that. The "1" came not here in front (left side), it got behind us (right side), and backed us up to make us into something recognized as having the power of "1" generated into nothing "0," right? I'm not the teacher that teaches something I don't know.

A Change In Rulership

Since we are really catching up to get going with the God of this world, the making of this world was 9,000 years into our calendar history. For many millions of years, we have been using a God wisdom for only 25,000 years. One man's wisdom rules for 25,000 years, after that we take Him down and put up another One to rule for 25,000 years. This is in accord with the circumference of our planet. Our planet earth is approximately 25,000 miles in circumference; not quite, but it's just a little being 25,000 miles and we use the word twenty-five thousand. Why do we not have it 25,000? Up at the top of our planet's rotation, which we call the Axis of the earth, we have at the axis end 23-1/2 degrees here to the circle of the Earth at the axis point. We have 23 1/2 degrees here.

We have also 23 Gods that write and justify history once every 25 years. They equal the poles degree here, they both measure the same 23 1/2 degrees. It's equal with the Gods that rule. There are really 24, but this number is used whenever they make up history; for instance, 23 Gods work and 1 does the calculating on the what they find.

The knowledge of the god of this world is being taught to us by the God of our world. Our God taught me that in the year "1" of our calendar time of 25,000 years, before we allowed another God to rule, the scientists said in the year 9,000, there will be a man born in that year, his name will be Yakub, and he will make a people who will live for 6,000 years. This 9,000 year is the beginning of the birth of the man you call white folks. They were made by this man - not created now, they were made by a man called Yakub. In the Bible, you call him Jacob. That's the same man. Jacob made, according to the Bible, 12 sons. Actually Jacob could not produce the 12 scientists. The 12 scientists were already here. Yakub did not come out making scientist; he was not that wise, though he was a wise man and a wise god. They called him the big head Yakub; meaning, he had lot's of brains.

So in this year of 9,000, this man comes out with a new civilization to take over us. He did not make no mistake. He took over us and it's hard for us to get away from him today. The Bible points this out accurately when it stated that they were to have dominion over all creatures, and all life. He has power over all the life regardless to what kind of life it is. He has had power over it and to use it as he pleases. He has power even over to the original man, the god, the original creator.

We, the originally created, was allowed to keep just a little, just a spark to keep from going out of the time. The Holy city Mecca had this little spark. This Holy City Mecca of Arabia, the original scientist didn't rule there, nor do they rule there today, but they're gradually easing in. They are white scientists who's ruling there. This is why Muhammad must destroy the old. You find in the history of Muhammad, he declared war on Mecca and he overcame them. So today I'm telling you that we are not going to fight to become the ruler under the same thing that the old world have ruled under. We're going to make our own Holy City.

You'll read in the Bible where it says that "I saw a new Jerusalem coming down from out of the throne of God - the new God, He has a new city; He doesn't let His new wisdom and His new kingdom be built upon an old foundation. (TOT 83-89)

"1" was already in the darkness, but nothing did not give it to us until the time brought it about. When the time brought it about, it was out there in our view as a revolving light; light revolving that's hidden there in the dark. We don't know how many trillions of years it was there, but it was there. With a fine atom of water out there, it made itself out of an atom of water that was found in the darkness of the universe. We could not see light emerging out of the space without water, because we can't produce life without water; therefore, it was some water out there in that darkened world of space. The space had produced life, but how long was that life out there before it produced us a form?

HOW TIME IS DEFINED

Calculate how long life takes to make 1 atom; this is the way it's understood. You and I being dead over here in America for so many years, and North America had been dead so many thousands, and the space over North America so many thousands, we can't calculate the actual time. By the instruments we have now, we can calculate time from estimating so long and so long. (TOT pg96)

What is time? The main thing we should know is what do we mean when we say time. We mean the distance of something that is taking us so long and so long to get to the other point, and if there are two points here to meet together, we count the time that we're going to try and meet between the two objects. It means that we have motion and motion makes time. We cannot have any reading of time until we set up motion. When motion is made, we calculate how long that motion takes to meet with another object. (TOT Pg120)

If you are going to be freed with the time, when did the time start? We cannot count on time unless we know when time began. Well, we don't want you coming out here saying we're talking on something we don't know nothing about. We know you and if we know you, we know you're talking about something. Yes. Time, when it started, it started way back, long ways back. It started when the first motion of an atom moved out of darkness. When that atom moves it's still time. That was, according to the teachings of Allah to me, was something around about 76 trillion years ago.

It was seventy-six trillion years ago, when that these things took place; "What things?" The motion of time. If I move my hand from one point to another point, you can say that I am making some kind of time, right? Because I'm doing this to make motion. To read the motion is to read how long it takes this hand to come from here to there or go back here - that's time. If I hold my hand still, it's making no time, nothing is going on; it's just a hand being held up.

As original black people here in North America, we have not been making time ourselves, and it's due to us not knowing how to make time. Time without a product is wasted time. We don't want that kind of time. We want time that will produce something. God is a production made by the action of time. We say then that we we're doing pretty well. As long as we have no production in the time, in that we are making, we have then to sit down - let somebody else come in and make time. We have been sitting down here now every since the slave master said, "you are free nigger." We sit down and ask him for something to do. (TOT pg121)

WHY WAS YAKUB ALLOWED TO RULE FOR SIX THOUSAND YEARS?

There was no God in his time that could prevent him. And it was necessary for us to learn from one that which was in us, in our midst, what he had that could rule others for six thousand years from 9,000 to 15,000. (TOT Pg131)

We could have ruled it all the while, but we wanted to try out our own self at being ruled by an inferior self. We have learned the inferior self will never make a world of peace and happiness for the people, because where there's disagreement, there is war. So, we the Muslims, we cannot, by nature, disagree on the subject of peace; we must have peace.

For 42 years, I have not done a thing in the world but studied and taught the word of God from His mouth. As it is written, we have the wisdom, divine from above this world. If we made the white man, grafted him out of ourselves - that's what happened, what can you tell us about it. There is not a black person in this house that will give the white race any credit for nothing, even their own wickedness, we had taken that spark out of the germ of the black man and made a man out of him so that he would be totally weak.

This is what the Gods of the black man did and we know him, I know the white people. I don't study his book to know him, I study him from the mouth of God. He have tried him up until the coming of God to show him up, as the Bible teaches him and you and the Holy Qur'an makes it a little more plainer and more direct than we have it in the Bible. We know the man and now you are being taught the utmost knowledge of the man to so you can see if whether the God of justice is unjust for destroying him today. This is why you're taught all the ends and odds of the mans very nature, for you to see that God is not doing you and me an injustice for destroying him today.

I want you to remember that you're making a fool out of yourself after God have now come and told us the truth of us, that we were the first ones and will be the last if there be a last. We don't see through the telescope of the scientist's wisdom, who don't see anything like an end to the Black man. You don't see what He was made out of, how He made Himself out of the material that comes last, the material of darkness comes last in the absence of light. He had taken that darkness, created Himself out of it: that which comes last so that there would never be a last to Himself. I'm very sorry over you that have pity and want to argue in favor of this world against a world that has no beginning and no ending. You can't calculate the hour, the day, or year of the creation; it's impossible. There was no writers there at the creation, because the creation had to make a writer out of himself first, and it was 6 trillion years after that when writing took place after his own creating.

I can be called a liar like all the others before me who were prophets and reformers of people. I don't expect you to give me the credit, but you will grow to do that. I could tell you today, but you will grow to it tomorrow. You're not able with the knowledge of this world to see into what I'm teaching you of a world that has no birth record to it, and have no ending. You can't create a thought that would correspond enough with the answer for you to tackle what I am teaching. I don't only defy you, but I defy the scientists of this world to attack what I am teaching you and prove that I am teaching you something false, just because it is not of his world or knowledge. I will give to any professor of the white mans world 10 thousand and 10 thousand times 10 thousand to prove one word I teach you as untrue. (TOT Pg. 288-289)

The Creator was created like me and He has the brains like me, that came with the creation. He could make those brains do just what I'm doing, if the Will and power of the 12 God-Scientists, Who confers, are with him. He can do just what I'm doing.

When Messenger is chosen, these 12 confer to see if whether or not he can do the job that he is now chosen for. The 12 Gods decide on all of this. We call them the 12 Great Scientists. This is why your ruler was made with 12 inches; it's after the number of the 12 Major Scientists, and without this, you cannot live. We call it a ruler when you have 12 square inches in it. (TOT Pg. 290)

The white man of America is not going to tell you how close his doom is. You will get proud and want to get even with him and that's right. We will believe that we are safe in trying to get even with him if we know the exact time, but he's not going to tell you. He's very close to his doom at this present hour, but he's not telling you. He want you to believe that he has yet a long future. He doesn't have one. The white man was not put on the earth to live forever. He's a new people among us who were made, not created. I want you to understand this, you and I are created people, but the white man is not the created people. He is a made people. You and I have a long, long old birth record. It's so long that we cannot tell you when we were born, but the white man, he is a man that our scientists made here recently 6,000 years ago, that's no more than talking 6 hours in the face of the ancient scientists. He's a very late man.

FROM THE DUST OF THE EARTH

He was made through grafting, from the dust of the earth. You must understand that this also has meaning. It doesn't mean a man was made actually from dust. This is science. The science means this: that he's made of the weaker and the lower character of the black man. God has given the knowledge to me to explain to you, that which you do not know and do not understand. (TOT Pg332-333)

In the 9,000th year, remember this, an enemy was produced to cause all of this trouble that has been coming to the black man for the last 6,000 years. We call him the white man instead of just saying devil, that's his real name - devil.

You go by his color, because you didn't know who was under the color. When Yakub was making him, he took the brown germ away from the black man and he took that brown germ, which was weaker than the black germ, and he looked into the possibility of making weakness rule strength. This is not out of the order of science. This weak germ in the black man was brown; so, Yakub said, "I will take the brown germ away from the black one, make a man out of the brown germ, give him my wisdom and my knowledge and then he can rule the black man from whom I took him. When the black man produces a man that is wiser than I, he will take me and my made-man away off the scene to never rise anymore."

He told his people, the white people, "When you have lived on this earth 6,000 years, don't try to run the 7th, because the in the 7,000th year, your brother coming from the East, He's going to eat you up. The white man or devil knows, this is why he goes over in the East and stocks his artillery over there to try to keep the man from coming out. He start working on him before he leaves, but they know he was going to do that. (TOT pg. 67-68)

THE BROTHER FROM THE EAST

According to the teachings to me of man's histories by Allah (God) in the Person of Master Fard Muhammad, praise is due to Him forever, the Great Mahdi and the Messiah that the world has been expected to come for the last two thousand years, has come and is going about His work as has been predicted that He would do.

He taught me for three years (night and day) on the histories of the two people, Black and white.

What he taught me verifies the teachings of the Bible and Holy Qur'an (if rightly understood) of the two people, Black and white.

He said that there was no birth record of the Black Man and therefore none can say how old the Black Nation is. As far back as, He said, a record has been kept, it dates 76 trillion years. These years (76 trillion) were divided into periods of six trillion years beyond 66 trillion years, which would make the said figure 76 be 78, instead of 76 trillion. I will not go into the finer details here. (OSHA PG. 96)

Jesus prophesied that: "You shall know, that which shall make you free," (John 8:32).

Today it has come to you and me from the Lord of the Worlds; it is a true knowledge of the Person of God and the

Devil. I am not responsible for your rejection. I am only responsible for the deliverance of it [the truth]. According to the Bible (John 8:42), Jesus said to this race that "If God (The God of righteous) were your father you would love me." In the 44th verse of the same Chapter, Jesus says: "You are of your father the devil." This proves beyond a shadow of a doubt that this race (Caucasian) cannot believe the Truth and especially Allah, His Prophets, and Islam. Being not of Allah, they cannot by nature love and obey Allah, His Prophets, nor the black Nations. Try teaching them of Allah and His religion, Islam; some of them will claim they never heard of Islam, to see how much you know of it, though their Father (Yakub), sometimes referred to as Adam, was once in Islam. (Foundation II, pg. 15)

The great arch deceivers (the white race) were taught by their father, Yakub, six thousand years ago, how to teach that God is a spirit (spook), and not a man. In the grafting of his people (the white race), Mr. Yakub taught them to contend with us over the reality of God by asking us of the whereabouts of that First (God). One Who created the heavens and the earth, and that, Yakub said, we cannot do. Well, we all know that there was a God in the beginning Who created all these things, and do know that he does not exist today, but we know again that from that God, the person of God continued until today in His people, and today a SUPREME ONE (God) has appeared among us with the same Infinite Wisdom to bring about a complete change.

This is He, of Whom, I preach and teach you to believe and obey. The devils call Him a Mystery God; yet, claim that He begot a son by Mary. They call on you and me to take this son of Mary for a God, who was a man before and after his death; yet they deny the coming of God to be a man. If Jesus was a son of God, what about Moses and the other prophets? Were they not His sons since they were His prophets?

The belief in a God other than man (a spirit), Allah has taught me, goes back into the millions of years - long before Yakub (the father of the devils) - because the knowledge of God was kept as a secret from the public. This is the first time that it has ever been revealed, and we, the poor rejected and despised people, are blessed to be the first of all the people of earth to receive this secret knowledge of God. If this people (the white race) would teach you the truth which has been revealed to me, they would be hastening their own doom - for they were not created to teach us the truth, but rather to teach us falsehood (the contrary to truth). (Foundation II, pg. 161)

Master Fard Muhammad is that Man Who has the wisdom or the knowledge of how the creation took place in the first place. What I mean: The creation of the universe, how it took place and how life and everything began down to six thousand years of Mr. Yakub and his world, and the knowledge of the world to come of which He Himself is the God. He's the God of the next world, He's planning it now. And that world He will set up, we see now gradually merging in. It will be forever. There will be no change in that one, because the scientists just don't see through the future of any other change. They don't have too much knowledge of His change, as He has not let go yet. He brought me up pretty close, but He has not as yet opened up. We have it in the Bible and Qur'an, that eyes have not seen nor ears have heard, and it has not entered into the heart of us what He will do, but we get little inklings. Now we are right in His time, and we do know that He is going to bring about a complete change.

As I've studied for thirty three years after His coming, I have learned my own self from study of His words, what He said to me. It won't be a world like we see today, nothing of the kind, not even to the Orthodox Muslims. Their world will not even be considered in His world. The Master Scientist or God, six thousand years ago, Mr. Yakub, did not make his world like our world, on the basis of our world. We have a completely new world and a new people in you. So will the great Mahdi, Master Fard Muhammad, He will also build a new world and a new people.

He did tell me how we would start making a change in a new people. Now just exactly what we will look like, I'm not to sure of that, but I believe He's going right back to the origin; He's going to make a better people. What I mean to say is a people that will be more stronger physically and they will be taught in such way that they can live much longer. Individuals will live probably a thousand or more years.

In teaching something the people on Mars, He often would refer to them, [while with me saying], "just think of them living twelve hundred of our Earth years." He said, and we're dying in less than a hundred years. It gives me an idea that He wants to make a people that will live a thousand or more years. This is also hinted at in the Bible and Qur'an, this teaching, and so now He intends to do just that. (HISTN)

THE BIRTH OF MR. YAKUB

Our 66 trillion years from the moon has proven a great and wise show of the original power, to build wonders in the heavens and earth. Six thousand years ago, or to be more exact 6,600 years ago, as Allah taught me, our nation gave birth to another God whose name was Yakub. He started studying the life germ of man to try making a new creation (new man) whom our twenty-four scientists had foretold 8,400 years before the birth of Mr. Yakub, and the Scientists were aware of his birth and work before he was born, as they are today of the intentions or ideas of the present world.

According to the word of Allah to me, "MR. Yakub was seen by the twenty-three Scientists of the black nation, over 15,000 years ago. They predicted that in the year 8,400 (that was in our calendar year before this world of the white race), this man (Yakub) would be born twenty miles from the present Holy City, Mecca, Arabia. And, that at the time of his birth, the satisfaction and dissatisfaction of the people would be: -- 70 per cent satisfied, 30 per cent dissatisfied.

And, that when this man is born, he will change civilization (the world), and produce a new race of people, who would rule the original black nation for 6,000 years (from the nine thousandth year to the fifteen thousandth year).

After that time, the original black nation would give birth to one, whose wisdom, Knowledge and power would be infinite. One, whom the world would recognize as being the greatest and mightiest God, since the creation of the universe. And, that He would destroy Yakub's world and restore the original nation, or ancient nation, into power to rule forever.

This mighty One, is known under many names. He has no equal. There never was one like Him. He is referred to in the Bible as God Almighty, and in some places as Jehovah, the God of Gods, and the Lord of Lords.

The Holy Qur-an refers to Him as Allah, the One God; beside Him, there is no God and there is none like Him; the Supreme Being; the mighty, the wise, the best knower; the light; the life giver; the Mahdi (this is He, Whom I have met and am missioned by).

He, also, referred to as the Christ, the second Jesus. The Son of Man, who is wise and is all-powerful. He knows how to reproduce the universe, and the people of His choice. He will remove and destroy the present, old warring wicked world of Yakub (the Caucasian world) and set up a world of peace and righteousness, out of the present so-called Negroes, who are rejected and despised by this world.

Mr. Yakub was, naturally, born out of the 30 per cent dissatisfied. As we know, wherever there is a longing or demand for a change, nature will produce that man, who will bring it about.

Allah taught me that the present percentage of dissatisfaction is 98 per cent near, 100 per cent, with the present ruling powers. This 100 per cent dissatisfied will bring about a 100 per cent change. Yakub did not bring about a 100 per cent change, but near (90 per cent). (MTTBM PG. 111)

YAKUB'S DETERMINED IDEA

Yakub was a playing with 2 pieces of steel one day and that's where he learned his own work by playing with 2 pieces of steel, watching their reaction - one attracting the other, that this steel would attract each other and come together and he looked at that reaction, he looked up to his uncle and said "Uncle when I get an old man..." that would now mean he was 6 years old. He was a man born a Black Scientist with a big head. He had a great big head, and they use to call him - so God told me - 'The big head Yakub,' because his big head contained so much wisdom or brains of wisdom that he was called 'the big head Yakub;' meaning he had a lot of wisdom. He told his Uncle, saying "...when I get an old man, I'm going to make a man that will rule you." His uncle looked at his nephew and said, "What will you make other than something that will cause bloodshed and mischief in the land?" He didn't tell his uncle, no, no. He knew that his Uncle was telling the truth. He said to his Uncle, "that's alright, I know what you do not know." Yakub did not tell him that he was wrong, because that was just the thing he was going to do; so he told him, "I know what you did not know and he did. (TOT 60)

He learned his future from playing with steel. It is steel and more steel that this made race (the white race), are still playing with. Steel has become the most useful of all metal for the people. What he really saw in playing with the two pieces of steel was the magnetic power of attraction. (MTTBM PG. 112)

"Muhammad who is this questioning Muhammad?" It's Gods, he was in the circle of Gods and that was God speaking to him, questioning him on his make of a man.

When Yakub got to be a grown man, he went out to start grafting the man that he decided he would make a ruler from us, think over that. He didn't go out to make him otherwise, but to get him out of us, because we had the right material in us for the making of such a man. You might be saying, "Well Muhammad, what kind of material?" Yakub, in his college, school in his day and time in the laboratory, took the germ of the black man and began experimenting on it to see what was in it. He saw in the black man's germ 2 people and one was weaker than the other. That germ was brown. So, he kept experimenting with it. He said, "Yes, I can separate this brown germ which is weaker than the other one - the black one - and I can make a man out of that germ. I will teach him my wisdom and then he will move [impose on] the black one until the black one produces a God greater than I. Yakub discovered that black germ that he would use to rule a superior man. It would take 6,000 years for the black germ to produce a white man that would have more wisdom than he, then he could erase my civilization and set his into orbit. (TOT60-61)

Unalike in the steel was making the steel to come together from the like, the attraction of the magnetic power drawing the 2 together. He saw his future right from the steel reacting. So today, for 6,000 years, they have been playing with steel. He saw that steel was a good piece of metal to murder with. He fashioned all kinds of weapons out of steel, they are using it today. They're clashing in war with steel. The first time they went to war with steel was with bow and arrows, shooting arrows. They began from stone to that. They learned to put a little steel on the end of the stone. So, today, he's learned to make a gun that could shoot steel into our bodies. He's been a murderer ever since he set foot on the face of the earth. (TOT83-89)

The piece attracting and drawing the other under its power. In this, he saw an unlike human being, made to attract others, who could, with the knowledge of tricks and lies, rule the original black man -- until that nation could produce one greater and capable of overcoming and making manifest his race of tricks and lies, with a nation of truth.

Yakub was the founder of unlike attracts and like repels, though Mr. Yakub was a member of the black nation. He began school at the age of four. He had an unusual sized head. When he had grown up, the others referred to him as the Big head Scientist.
At the age of 18 he had finished all of the colleges and universities of his nation, and was seen preaching on the streets of Mecca, making converts. He made such impressions on the people, that many began following him.

THE SCIENCE OF THE GODS

He learned from studying the germ of the black man, under the microscope, that there were two people in him, and that one was black, the other brown.

He said if he could successfully separate the one from the other, he could graft the brown germ into its last stage, which would be white. With his wisdom, he could make the white, which he discovered was the weaker of the black germ (which would be unalike) rule the black nation for a time (until a greater one than Yakub was born). (MTTBM PG. 113)

According the words to us from Almighty God Allah, in the Person of Master Fard Muhammad, to Whom Praise is due forever, Who has now made manifest the secret of the Gods. This is the secret of the Gods that we are discussing, and not that of mere human beings. In taking a god from a God, the god that came from the original God or race of people or nation or aboriginal, is easy to make into what the name means, devil, because by grafting the very nature, the entire human being, is more fit for the teaching of grafting or law, than something other than that which he came from.

Now as we see in fruit, fowl and in the grafting of animals and what not, that the product always become lesser and lesser as we continue to graft. Now we come to the white man and the black man: Here we have a black man, who is the original man, of whom, we can't locate his birth; he's the aboriginal human being of the Earth. Now from him he has produced brown, yellow, red, and white, through the master wisdom of grafting of the God, Yakub. In this work of Mr. Yakub, taking the white man out into, what we would say today, the fifth stage of the family, he became the fifth man, and the most distinct of all. He's a new man to every one of us, a new race of people with a nature and wisdom that is absolutely new and contrary to the aboriginal, the black man. Before Mr. Yakub, they had never reached this particular point of Unalike attraction or Like repelling, but in his producing a people, this is the base on which he set. This was his work which produced an Unlike people, and this Unalike people (white race) would be able to attract the Like (Black Nation). In this way, he had perfected an idea of how to rule a Like. Unalike attract and Like repel; therefore, being able to produce this Unalike, it would attract a Like and will be able to rule a Like until a Like produces something to prevent a Like from being attracted by an Unalike.

This refers to the Gods. We are into what the scientists refer to as the wisdom of God, and how Their ideas work, and how Their ideas come to perfection.

Here we have a man produced from the Original man, and that man we call the white man. The white man, we say, is the devil. Why, because of his physical weakness, caused by coming into being from the original man, or aboriginal people of the Earth. As a result of their wise mastermind father, the God that made him, he is able, with that weak physical form taken from the original form or original man, rule the strong man, which is mentioned so much among the scholars and scientist that write history and discuss this point in their language, which the masses cannot understand.

This made it easy to teach the man a wisdom contrary and opposed to the wisdom of right, because Yakub was not after producing a man that was of the same or would think the same, like the one from whom he was taken. He had to make his world so attractive, that he would attract the minds of the world that he came out of until that world produced a man that would overcome him in wisdom. Ours world is superior in wisdom and would make His [the new God's] people equally superior by teaching them the origin of everything that took place and then the white race would have no more attraction upon the people that they came from.

After they learned the origin of their creation and how it came about and how their God gave them the wisdom and know-how to rule that which they came out of; we could say, in words, the original overtakes himself again. [The New God would] starts all over again to manufacture or to produce another man as we have this in history and prophesy, that the God at the end of the world of the white man will produce a new heaven or new civilization. We say heaven or kingdom on Earth, but it only means people, a new people and they will have a new kind of a wisdom that is superior to the present. (HISTN)

THE ORIGIN OF RACES

According to what Master Fard Muhammad taught me on the origin of the races, they number around four, and from these four races of people, they have produced many different types of people, but they are not independent in their beginning, because they came from one. We have various colored people all over the Earth from brown to white; however, we're not all the same color. As a result of intermixing with such colors as black, brown, yellow, red and white, this has produced many other various colors. The origin of it, according to the teachings of Master Fard Muhammad to me, was from a Scientist or God - we see him as a God. Back six thousand years ago, a master grafting work on the human being started to produce a new civilization, or a new race of people from the original race of people or aboriginal people.

This man, Yakub, discovered in the germ of the black man, that he had two people in him, and had learned through study and experimenting on the germ that this second germ could produce a powerful people that would be able to rule that which they came from. They would be able to rule for around six thousand years or until the father or the aboriginal people produce One superior to his man. He had taken, through experimental work on the germ of man, a people of what we call today a white race, but before he produced that white race, he produced a brown race, then he produced a yellow race and so on. There lies his first grafting of the black man according to the teachings of Master Fard Muhammad, Whom we see and know today as being God in person.

From the grafting in it's first stage, he (Yakub) had a brown race of people from a black people, and it taken him, according to the teachings of God, two hundred years to produce that brown race, and he kept up the process of killing off the brown or the darker one and marrying the lighter one with the lighter one, and in another two hundred years, he had a yellow race of people.
In this length of time, these brown people were spreading and migrating over the earth to find them a home to themselves.

And so, when the yellow race was produced, they too started migrating over the Earth, and from the yellow race, about two hundred years, this grafting kept in process. They were on the same Island, according to the teachings of Almighty God that were taught to me, by Allah (God), Who came in the Person of Master Fard Muhammad, to Whom Praises is due forever. During that two hundred years of keeping up birth control laws, this is what he established. He established it on this Island in his lab where he was working. We call it a human lab to produce his man. And from this yellow race he had a white race two hundred years later from them. And this was the end of his work. This was the man he was trying to get to: a white race.

That concluded the total time of his grafting the white out of black in six hundred years, and this figure tallies with the Bible's teaching of the man being created in six days. The days here mean a hundred years each, six hundred years. And this also tallies with the creation of the universe: six.

The whole entire universe, according to the Qur'an was also created in six periods of time. And so Mr. Yakub, the Mighty Scientist of that time, produced his man on six, because it tallied with the creation of the universe and it was according to this number that he could do his work. They would be masters, gods to rule the Earth, the people and everything of life, for six thousand years. At the end of six thousand years, the aboriginal people will have produced another One mightier than Yakub in wisdom and knowledge and He would be equal with that One that created the heavens and the Earth. And He would have the power and the infinite wisdom to make His Word Be as the first One. This is the Man we have before us today, to Whom praise is due.

According to the way that Almighty God taught me, in the Person of Master Fard Muhammad, that in grafting anything, even fruit, or the life of animals like the birds, or fowl, all of these things can be grafted to make another self out of them. So you have gotten just this: You have made many things from the aboriginal. And today we see much of your world that you have produced through grafting. And you're a master at that. I must say, you're a master at grafting, because the father, Mr. Yakub, was a master.

According to what I have been taught of God, as you know, that I don't know anything and I didn't know anything before this Master Mind came to me or the Wise Man Who I teaches to be God in Person. I believe I'm right according to the scripture, and I know I'm right according to what He has said, and how it corresponds with the prophetic sayings of the prophets concerning him, or one coming in this time.

In grafting, lets get back there to answer that question. In grafting a life from a life that is original, that which we graft from the original, regardless to what life it is, it is weaker than that which it was grafted from. We are reducing the power of that original when we graft from it. When we graft fruit, the fruit that we graft is not the equal to that which we graft it out of. And so it is when it comes to life, whether it's birds, whether it's a beast, animals, or a man.

In cats, such as the cat family, all the grafted of it's family is lesser in power than the original lion, and that which is grafted from the original lion is not to be trusted to much as with leopards and in the other cats from the lion family. These cats from the lion family are more dangerous and cannot be trusted like we can this original lion. The original lion, we can take him, make him lie around us even at an old age, and we can still trust him, but we can't trust this leopard and these other cats, because they have, by nature, something evilly wicked that came through the grafting. He comes upon us while we're not looking at him and before we know it, he'll destroyed us. So we don't trust him, but we will trust the original lion. Get them when he's a little cub, and we can bring him up like he's any other house cat, walk around and play with him till he's an old man. So it is with the human being. (HISTN)

Any live germ grafted from the original is devil. Satan, on the other hand, has a wider spread, we would say, than the devil. We can see a devil cat out there, a devil animal, or a devil person. We could call most any grafted thing a devil, you see, but actually when we say Satan, we mean a man or a people whose wickedness is not confined to themselves, it spreads, and others are affected by their wickedness.

Now we can take this question up from just what we said here in the grafting and through the grafting of this people. They were being separated all the while. The brown was separated from the yellow, and yellow separated from the white. So, separation was the answer to the whole process of the creation, and the whole universe; everything is separated. There's nothing put in the sphere of another one. Everything is independent within it's own sphere. So, since the white man brought us out of our original land and people and has kept us here without taking us back or without the knowledge of how to get back for the past four hundred years, this knowledge God has been brought to us, and the answer is now only separation. We must go back to our own.

[The whites] came here from Europe. You probably will have to return to your home in order to fulfill that which is written of the scripture by the prophets who put it in there. I think I read this somewhere in the Bible where it says at that time, every man will go to his own people and country, meaning at the end of the rule of the white people, who will then separate themselves and every man will go like they did in the beginning. At the beginning they were separated, the black went to itself, the brown went to itself, yellow went to itself, white went to itself.

The color or the races are people, and Islam is religion. When we say the people of Islam, we mean of that religion or that duty to God, and when we refer to the color of the various people of Earth, we are referring to races and nations as they're called: Such as brown, yellow, red colors. They're all from the same source or the same father, the black man; he has that in him. The white color, it's a color, and according to the teachings of Master Fard Muhammad, to Whom Praise is due, God in Person, we call white the fifth color from black and the final color of black. It's not anymore part of the other four. It's completely to itself. It's independent now to itself.
Its root is from the first, the black , but it has now been taken completely from that and within itself it's independent. (HISTN)

JACOB WRESTLES WITH THE ANGEL

This new idea put Yakub to work finding the necessary converts to begin grafting his new race of people. He began by teaching Islam, with promises of luxury to those who would believe and follow him.

As Mr. Yakub continued to preach for converts, he told his people that he would make the others work for them. (This promise came to pass). Naturally, there are always some people around who would like to have others do their work. Those are the ones who feel for Mr. Yakub's teaching, 100 per cent.

As he made converts in and around the Holy City of Mecca, persecutions set in. The authorities became afraid of such powerful teachings, with promises of luxury and making slaves of others. As they began making arrests of those who believed the teaching, the officers would go back and find, to their surprise, others still teaching and believing it.

Finally they arrested Mr. Yakub. But, it only increased the teachings. They kept persecuting and arresting Yakub's followers until they filled all the jails.

The officers finally reported to the King that there was no room to put a prisoner in -- if arrested. All the jails are filled; and, when we go out into the streets, we find them still teaching. What shall we do with them? The King questioned the officers on just what the teachings were; and of the name of the leader.

The officers gave the King the answers to everything. The King said: This is not the name of that man. On entering the prison, the King was shown Yakub's cell. Wa-Alaikum. The King said: So you are Mr. Yakub? He said: Yes, I am. The King said: Yakub, I have come to see if we could work out some agreement that would bring about an end to this trouble. What would you suggest?

Mr. Yakub told the King: If you give me and my followers everything to start civilization as you have, and furnish us with money and other necessities of life for twenty years, I will take my followers and we will go from you.

The King was pleased with the suggestion or condition made by Yakub, and agreed to take care of them for twenty years, until Yakub's followers were able to go for themselves.

After learning who Mr. Yakub was, they all were afraid of him, and were glad to make almost any agreement with him and his followers.

This history or future of Mr. Yakub and his people was in the Nation's Book, by the writers (23 Scientists) of our history, 8,400 years before his birth. So, the Government began to make preparation for the exiling of Mr. Yakub and his followers. The King ordered everyone rounded up who was a believer in Mr. Yakub. They took them to the seaport and loaded them on ships.

After rounding them all up into ships, they numbered 59,999. Yakub made 60,000. Their ships sailed out to an Isle in the Aegean Sea called Pelan (Bible Patmos). After they were loaded into the ships, Mr. Yakub examined each of them to see if they were 100 per cent with him; and to see if they were all healthy and productive people. If not, he would throw them off. Some were found to be unfit and overboard they went.

When they arrived at the Isle, Mr. Yakub said to them: See how they (the Holy people) have cast us out. Now -- if you will choose me to be your King, I will teach you how to go back and rule them all.

Of course, they had already chosen Yakub to be their King at the very start. So, Yakub chose doctors, ministers, nurses and a cremator for his top laborers. He called these laborers together and told them his plan for making a new people, who would rule for 6,000 years.

He called the doctor first and said: Doctor, let all the people come to you who want to marry; and if there come to you two real black ones, take a needle and get a little of their blood and go into your room and pretend to be examining it, to see whether their blood would mix. Then, come and tell them that they will each have to find another mate, because their blood does not mix. (It was the aim of Yakub to get rid of the black and he did.) Give them a certificate to take to the minister, warning the minister against marrying the couple because their blood does not mix. When there comes to you two browner ones, take a pretended blood test of them; but, give them a certificate saying that they are eligible to marry.

Mr. Yakub's charge to his laborers was very strict -- death if one disobeyed. They didn't know what Yakub had in mind until they were given their labor to do. He made his laborers, from the chief to the least, liars. The doctor lied about the blood of the two black people who wanted to marry, that it did not mix.
The brown and black could not be married (brown only). The doctors of today hold the same position over the people. You go to them to get a blood test to see if you are fit to be married.

Today, they say it is done to see if there are any contagious germs in the blood. I wish that they would enforce such a law today (keep the white from mixing with black -- just the opposite). Perhaps we could remain black and not be disgraced by a mixture of all colors.

In the days of Yakub's grafting of the present white race, a new and unalike race among the black nation for 600 years, his law was -- that they should not allow the birth of a black baby in their family, but the white (devil) should mix their blood with the black nation, in order to help destroy black; but, they should not allow the black to mix with their blood.

His aim was to kill and destroy the black nation. He ordered the nurses to kill all black babies that were born among his people, by pricking the brains with a sharp needle as soon as the black child's head is out of the mother.

If the mother is alert (watching the nurse), then the nurse would lie and fool the mother to get possession of child to murder it, by saying that she (mother) gave birth to an Angel child. And that she (the nurse) would like to take the baby to heaven, so when the mother dies, she would have a room with her child in heaven, for her baby was an angel.

This is the beginning of the first lie or liar; and, it was so that the nurse would take the black baby away on this falsehood and claim that they were taking the poor black baby to heaven. As Yakub had taught them, they would feed it to wild beasts and if they did not have a wild beast to feed the black babies to, Yakub told the nurses to give it to the cremator to burn.

Mr. Yakub warned the laborers, from the doctor down to this cremator, that if anyone of them failed to carry out his orders, off go their heads.

When there was a birth of a brown baby, the nurse would come and make much ado over it, and, would tell the mother that she had given birth to a holy child and that she would nurse it for the next six weeks, for her child was going to be a great man (that is when it was a boy baby). (MTTBM PG. 117)

This race you call the Japanese people, that's the brown race. Okay, the next is a yellow race. Now from yellow, this other race that you see here is white, this number 1 here. (I shouldn't have balled it up there so close to that, but I was saying the brown race was the first one to come out from grafting.)

The first man here was black, then when he started grafting out of him to get to white, he was grafting for 200 years. For 200 years, it took him that length of time to get the brown baby out of black. Yakub found the brown germ in us while studying in college. He discovered in the germ of the black man that he had a brown germ in him or that he had 2 - we will say 2 babies - one brown and one black. He kept looking at that germ - brown - and he discovered that he could graft it out of the black man, take that brown germ, keep it to itself and keep grafting it until he took the brown germ out and replace it with a more different color, if he could just keep taking the more lighter brown one that came, and keep it to itself, then kill off the brown one, and he did that for 600 years.

All through grafting, this was done, and every time he'd graft one, it was weaker than the other: The brown baby was weaker than his father, the black man; the brown baby was stronger than the yellow baby, and the yellow baby, who produced the white baby, was a little stronger than the white, base on the nearness back to the father of [all the races] black man. The little yellow baby became kind of evil and wicked, like the next brother, white. This is the Chinese that I'm talking about. The China man in near equal with white folks. He loves to fight, but he's braver than whites, because he's back more closer to the father than the white. The father of race is the black man. And these people go and mix with many, and by mixing with many, they produce many different colors; thus, many different races. So I want you, my beloved readers, to remember that the white man didn't come from the same God that you came from. (TOT 83-89)

After the first 200 years, Mr. Yakub had done away with the black babies, and all were brown. After another 200 years, he had all yellow or red, which was 400 years after being on Pelan. Another 200 years, which brings us to the six hundredth year, Mr. Yakub had an all-pale white race of people on this Isle.

Of course, Mr. Yakub lived but 150 years; but, his ideas continued in practice. He gave his people guidance in the form of literature. What they should do and how to do it (how to rule the black nation). He said to them: When you become unalike (white), you may return to the Holy Land and people, from whom you were exiled.

The Yakub made devils were really pale white, with really blue eyes; which we think are the ugliest of colors for a human eye. They were called Caucasian -- which means, according to some of the Arab scholars, One whose evil effect is not confined to one's self alone, but affects others.

There was no good taught to them while on the Island. By teaching the nurses to kill the black baby and save the brown baby, so as to graft the white out of it; by lying to the black mother of the baby, this lie was born into the very nature of the white baby; and, murder for the black people also born in them -- or made by nature a liar and murderer.

The black nation is only fooling themselves to take the Caucasian race otherwise. This is what Jesus learned to their history, before He gave up His work of trying to convert the jews or white race to the religion of Islam.

And, the same knowledge of them was given to Muhammad by the Imams (or scientists) of Mecca. That is why the war of the Muslims against them came to a stop.

Muhammad was told that he could not reform the devils and that the race had 1,400 more years to live; the only way to make righteous people (Muslims) out of them was to graft them back into the black nation.

This grieved Muhammad so much that it caused him heart trouble until his death (age sixty-two and one half years). The old scientists used to laugh at Muhammad for thinking that he could convert them (the devils) to Islam. This hurt his heart.

Mr. Yakub taught his made devils on Pelan: That -- when you go back to the holy black nation, rent a room in their homes. Teach your wives to go out the next morning around the neighbors of the people, and tell that you heard her talking about them last night.

When you have gotten them fighting and killing each other, then ask them to let you help settle their disputes, and restore peace among them. If they agree, then you will be able to rule them both. This method the white race practices on the black nation, the world over. They upset their peace by putting one against the other, and then rule them after dividing them.

This is the reason why the American so-called Negroes can never agree on unity among themselves, which would put them on top overnight. The devils keep them divided by paid informers from among themselves. They keep such fools among us. But, the real truth of the devils sometimes converts the informers and brings them over to us as true believers. We don't bother about killing them, as I am not teaching that which I want to be kept as a secret, but that which the world has not known and should know.

After Yakub's devils were among the Holy people of Islam (the black nation) for six months, they had our people at war with each other. The holy people were unable to understand, just why they could not get along in peace with each other, until they took the matter to the King.

The King told the holy people of the black nation that the trouble they were having was caused by the white devils in their midst, and that there would be no peace among them until they drove these white made devils from among them.

The holy people prepared to drive the devils out from among them. The King said: Gather every one of the devils up and strip them of our costume. Put an apron on them to hide their nakedness. Take all literature from them and take them by way of the desert. Send a caravan, armed with rifles, to keep the devils going westward. Don't allow one of them to turn back; and, if they are lucky enough to get across the Arabian Desert, let them go into the hills of West Asia, the place they now call Europe.

CROSSING THE BURNING SAND

Yakub's made devils were driven out of Paradise, into the hills of West Asia (Europe), and stripped of everything but the language. They walked across that hot, sandy desert, into the land where long years of both trouble and joy awaited them; but -- they finally made it. (Not all: many died in the desert.)

Once there, they were roped in, to keep them out of Paradise. To make sure, the Muslims, who lived along the borders of East and West Asia, were ordered to patrol the border to keep Yakub's devils in West Asia (now called Europe), so that the original nation of black man could live in peace; and the devils could be alone to themselves, to do as they pleased, as long as they didn't try crossing the East border.

The soldiers patrolled the border armed with swords, to prevent the devils from crossing. This went on for 2,000 years. After that time, Musa (Moses) was born: the man whom Allah would send to these exiled devils to bring them again into the light of civilization. Before we take up this first 2,000 years of the devils exiled on our planet, let us not lose sight of what and how they were made, and of the god who made them, Mr. Yakub.
Since we have learned that Mr. Yakub was an original man (black) the ignorant of our people may say: If Yakub was a black man and the father of the devils, then he was a devil.

That is like one saying The horse is as much a mule as the mule.

Or, that an orange or lemon is as much grapefruit as the grapefruit: because the grapefruit is grafted from the orange and lemon. They are not alike because the grafted is no longer original.

Just what have we learned, or rather are learning from this divine revelation of our enemies, the devils? Answer: We are learning the truth, which has been kept a secret for 6,000 years concerning the white race, who have deceived us. We learn what is meant by the Bible's symbolic teachings: that they were made from dust.

This only tends to convey the idea that they were created from nothing; which means the low and humble origin of such creation.

Again, we learn who the Bible (Genesis 1:16) is referring to in the saying: Let us make man. This US was fifty-nine thousand, nine hundred and ninety-nine (59,999) black men and women; making or grafting them into the likeness or image of the original man.

Now that they are the same, but have the ways of a human being they are referred to as mankind -- not the real original man, but a being made like the original in the sense of human beings.
The Holy Qur-an throws a great light on the truth of the creation of this pale, white race of devils. O mankind,

surely we have created you from a male and a female (Chap. 49:15). This makes it very easy to understand to whom it is referring. What mankind? Surely we created man from sperm mixed (with ovum) to try him, so we have made him hearing and seeing (Chap. 76:2).
Inasmuch as these chapters have a further reference to the spiritual creation of the Last Messenger, it is equally true that they refer to the physical creation of the white race. In another place, the Holy Qur-an says: We have created man, and now he is an open disputer.

Yakub's race of devils were exiled in the hills and caves of West Asia (now called Europe). They were without anything to start civilization and became savages. They remained in such condition for 2,000 years--no guide or literature.

They lost all knowledge of civilization. The Lord, God of Islam, taught me that some of them tried to graft themselves back into the black nation, but they had nothing to go by. A few were lucky enough to make a start, and got as far as what you call the gorilla. In fact, all of the monkey family are from this 2,000 year history of the white race in Europe.

Being deprived of divine guidance for their disobedience, the making of mischief and causing bloodshed in the holy nation of the original black people by lies, they became so savage that they lost all their sense of shame.
They started going nude as they are doing today (and leading the so-called Negroes into the very acts).

They became shameless. In the winter they wore animal skins for clothes and grew hair all over their bodies and faces like all the other wild animals.

In those days, they made their homes in the caves on hillsides. There is a whole chapter devoted to them in the Holy Qur'an. They had it very hard, trying to save themselves from being destroyed by wild beasts which were plentiful at that time in Europe.

Being without a guide, they started walking on their hands and feet like all animals; and, learned to climb trees as well as any of the animals. At night, they would climb up into trees, carrying large stones and clubs, to fight the wild beasts that would come prowling around at night, to keep them from eating their families.

Their next and best weapons were the dogs. They tamed some of these dogs to live in the caves with their families, to help protect them from the wild beasts. After a time, the dog held a high place among the family because of his fearlessness to attack the enemies of his master. Today, the dog is still loved by the white race and is given more justice than the so-called Negroes, and, is called the white man's best friend. This comes from the cave days.

AND THE SERPENT WAS LIFTED UP

After 2,000 years of living as a savage, Allah raised up Musa (Moses) to bring the white race again into civilization: to take their place as rulers, as Yakub has intended for them. Musa (Moses) became their God and leader. He brought them out of the caves; taught them to believe in Allah; taught them to wear clothes; how to cook their food; how to season it with salt; what beef they should kill and eat; and, how to use fire for their service. Moses taught them against putting the female cow under burden.

He established for them Friday as the day to eat fish, and not to eat meat (beef) on that day. And, fish is the main menu on Fridays in many of the whites' homes today.

They were so evil (savage) that Moses had to build a ring of fire around him at night; and, he would sleep in the center of the ring to keep the devils from harming him. They were afraid of fire, and are still afraid of fire.

Allah said that: One day, Moses told them he was going to have fish come up from the sea that so that tomorrow we will have some fish.

On the next day, the fish were there. Moses had a boat load sent up from Egypt. Moses said: See! The sea came up last night and brought us some fish. One of the savages was a little smart and he said to Moses: Where is the water?

From then on, Moses recognized the fact that he could not say just anything to them. He had a hard time trying to civilize them. Once they gave Moses so much trouble that he took a few sticks of dynamite, went up on the mountainside, placed them into the group, and went back to get those who were giving him the most trouble.

He said to them: Stand there on the edge of this mountain and you will hear the voice of God. They stood there about 300 in number. Moses set the fuse off and it killed all of them.

The Imams got after Moses for performing this trick on the devils. Moses said to the Imams: If you only knew how much trouble these devils give me, you would do as I do. Moses taught the devils that if they would follow him and obey him, Allah would give them a place among the holy people. Most of them believed Moses, just to get out of the caves.

The Imams recognized the tremendous job Musa (Moses) had, trying to civilize the savages. These enemies of the righteous black nation of earth now had to take the place as the rulers and conquerors of the earth. The devils were given the knowledge and power to bring every living thing, regardless of its kind of life, into subjection.

Let Us Make Man In Our Image

And God said: Let us make man in our image, after our likeness: Let them have dominion over the fish of the sea; and over the fowl of the air; and over the cattle, and over all the earth; and over every creeping thing that creepeth upon the earth: and God said unto them: Be fruitful and multiply; and replenish the earth, and subdue it (Gen. 1:26,28).
The above was all necessary if the devils were to rule as a God of the world. They must conquer, and bring into subjection, all life upon the earth -- not land life alone, but they must subdue the sea and the like therein -- master everything, until a greater master of God comes, which would mean the end of their power over the life of our earth.

We all bear witness that the scripture quoted above refers to the Caucasian race. They are the only people who answer that description and word for the past 4,000 years.

They have subdued the people and most every kind of living thing upon the earth. God has blessed them to exercise all their knowledge, and blessed them with guides (prophets) from among our own people; and, with the rain and seasons of the earth.

Today, their wealth is great upon the earth. Their sciences of worldly good have sent them, not only after the wealth of other than their own people, but even after the lives and property of their own kind. They have tired to re-people (replenish) the earth with their own kind, by skillfully killing off the black man and mixing their blood into the black woman.

But, the job is too big for them to ever conquer. The black nation, including its other three colors, brown, red and yellow, outnumber the Caucasian race, eleven to one.

God created them in His image (Gen. 1:27). They are in the image and likeness of a human being (black man), but are altogether different kind of human being than that of the black human beings.

Their pale white skin; their blue eyes (even disliked by themselves) tells any Black man or woman, that in those blue and green eyes, there just can't be any sincere love and friendship for them. They are unlike and we are like. Like repels -- unlike attracts. The very characteristics of black and white are so very different. (MTTBM PG 114-122)

The Truth is never wanted by false believers. The people, who have not known the Truth, must be taught it before Judging them. If a liar has deceived the people, then Truth will make manifest the Liar. Yakub's race of devils were made unlike the original Black Nation, for the purpose of being able to attract the original Nation to follow them. It stands true - White attracts Black and Black attracts White; but, where the attraction differs is, the Black - and especially the so-called Negroes - love the unalike (the

White Race), while the unalike, by nature, can't love the so-called Negroes.

As you have learned, they were made to hate and kill off the Black. They have killed on an average of 100,000,000 to every one thousand years that they have been on our planet; (600 million) and, hope to carry 97% of the so-called Negroes here in America to their doom, and two-thirds of the entire Black population of the Earth. Maybe a thorough knowledge of their enemies will reduce this figure. (FOUNDATION II 149)

The devil has deceived us about the true God. There's nothing to what he has taught us to believe in as God. It naturally does not exist - a spook for a god, and a dead prophet of two thousand years ago, whom they want you to believe is somewhere in a heaven, alive. Nothing could be worse, and the people believe it without any proof. I would like to make you prove such a false doctrine or suffer the consequences. There is no such thing as a Heaven nor Hell for one to go to after death. This is one of the first lies Yakub taught his made devils to teach to the people (you and me. Even to this day, there are millions of people believing such a lie. These are two conditions in our life: Heaven and Hell - not beyond the grave. DEATH SETTLES IT ALL! (Foundation II pg. 187)

ACCORDING to the Bible (Gen. 3:20-24), Adam and his wife were the first parents of all people (white race only) and the first sinners. According to the Word of Allah, he was driven from the Garden of Paradise into the hills and caves of West Asia, as they now call it, "Europe," to live his evil life in the West and not in the Holy Land of the East. "Therefore, the Lord God sent him (Adam) forth from the Garden of Eden, to till the ground from whence he was taken. So he drove out the man; and He placed at the East of the Garden of Eden Cherubims (Muslim guards) and a flaming sword, which turned every way to keep (the devils out) the way of the tree of life (the nation of Islam)." The sword of Islam prevented the Adamic race from crossing the border of Europe and Asia to make trouble among the Muslims for 2,000 years after they were driven out of the Holy Land and people, for their mischief-making, lying and disturbing the peace of the righteous nation of Islam.

The Holy Qur'an says: "But the devil made them both fall from it, and caused them to depart from that (state) in which they were; and we said: Get forth, some of you being the enemies of others, and there is for you in the earth an abode and a provision for a time."

(The time here refers to the limited time of the Adamic race. The time is 6,000 years). According to the above verse (2:36), they were driven out, because they were the enemies of the people of the Garden, in these words: "Get forth, some of you being the enemies of others." The "others" cannot refer to any others than the people of the Garden (the Muslims).

THE ADAMIC RACE is still the enemy of the Muslims (the black man). Nevertheless, Allah did not deprive the Adamic race of right guidance through His prophets, whom they persecuted and killed. The Adamic race's (the white race) history is a proof that they are the enemies of God and the righteous, for they never did sincerely accept a prophet of God. Can they now claim to be the chosen race of God? Where is their proof? Is it because they were allowed to rule us for 6,000 years? If they are the chosen race of God, why would God limit their time of rule? Why did God send His prophets to warn them that He was going to destroy them? Holy Qur'an (7:14):

"He said (the devil), respite me until the day when they are raised up."

Those that are referred to as being "raised up" refer to the resurrection of the black man into the knowledge of the white race as being the devils, the enemies of Allah (God) and the black nation.

"He said (the devil), as Thou hast caused me to remain disappointed, I will certainly lie in wait for them in Thy straight path." (Holy Qur'an 7:166).

What Allah disappointed the devils in was the limiting of their rule over the nations and making them manifest to the world of black men that they are the enemies and great deceivers of the righteous.

The white race is not, and never will be, the chosen people of Allah (God). They are the chosen people of their father (Yakub, the devil). (Foundation II, pg. 225)

THEY ARE bold and shyless, they are pig eaters, alcohol drinkers which help make them shameless, they go indecently dressed, according to the Holy Qur'an (7:17,20,21,26.) And it (Qur'an) warns us of the shame with which the devils deceived our fathers in "Paradise" 6,000 years ago. They (devils) led them out on an Island into the Agean Sea, stripped them of both the clothing that covered their shame and the wisdom and knowledge of themselves and God. And today they (the devils) are playing the same trick on you and me and the whole of the Black Nations.

The Holy Qur'an reads: "O children of Adam, we have indeed sent down to you clothing to cover your shame, and clothing for beauty, and clothing that guards against evil - that is the beast." This is of the Messages of Allah that they may be mindful, not the devils (the white race) seduce you, as he expelled your parents from the Garden, pulling off from them their clothing that he might show them their shame. He sees you, he as well as his host from whence you see them not, (electronic devices make this possible) surely we have made the devils to be the friends of the disbelievers - (7:26,27)." (Foundation I, pg. 24)

Allah told me that six thousand years ago, this same people's fathers (white race) broke the peace of the righteous in the gardens of Eden (the place that is known as old Persia). They called the truth of God a lie and made lies the truth.

They said to the people of the Garden, according to the Bible (Gen. 3:4-6),

"And the serpent (a name used according to the evil, deceiving characteristics of the Caucasian race) said unto the woman, you shall not surely die; for God doth know that in the day you eat thereof, then your eyes shall be opened, and you shall be as gods..."

The serpent lied, because they did die. And the deceiver (serpent) was driven out of the garden into the wilderness of the earth to build a wicked kingdom of evil to be destroyed on the coming of God.

They (the white race) are playing the same trick on the black nation today as they did in the days of Adam. They shall suffer eternal expulsion from this earth in a lake of fire (Rev. 20:10). Can they enjoy peace? After they were cast out of the garden, according to the Holy Qur'an; 7:16, "because thou hast thrown me out of the way, Lo, I will lie in wait for them on the straight way." As he had deceived Adam and his wife, he now declares that he will deceive the righteous in their straight path in the days of the resurrection and judgement of his evil world. He swore to them in another place that he would lead them to a tree of immorality and a kingdom that decays not. This kind of teaching is found in the teachings of Christianity. It is a very clever way of deceiving the black people of America, for here, Satan represents himself as an angel of light. They paint a picture of lies of "beyond the grave," when they know there is no life or communication with the dead. All ceases to be life after death. This is universally known.
(Foundation I, pg. 124)

The awakening of the Arabs nearly fourteen hundred years ago to the ancient Truth (Islam) (not a new Truth) of Allah by Muhammad and his work, was typical of what will be done today. He was opposed by the Arabs for a while. Although the Arab Nation and their country are the birthplace of the great prophets and Scriptures, and from there, prophets are sent throughout the world from the time of Adam until today. The Arab history cannot be compared with the history of the so-called Negroes who have never had a Divine prophet nor a Scripture!

According to the past histories of the major prophets, one comes 2000 years until the end of the World of Sin. Moses came exactly two thousand years after Yakub (the God and Maker of the evil Caucasian race). Jesus came two thousand years after Moses, and the last prophet came two thousand years after Jesus, fulfilling much of the histories of the Prophets before him; especially Moses, David, Jesus and Muhammad. The man Allah (God) raised up from among the American so-called Negroes in the West will unite his people to Islam with the guidance of Allah with a book of Scripture for his people prepared and written by the fingers of Allah (God). (Foundation I, pg. 186)

The wise white scientists know it as well as I, and hope to prevent it, but they are out numbered by their wicked wise. God created two kinds of everything, so it is with man. He didn't intend for them (Negro and white) to mix, and has set a day of reckoning for those who willfully, and knowingly break His natural law in which He hast created man. It is their purpose to try and make Allah (God a liar, so as to deceive the blind, deaf, and dumb so-called Negroes, as the devil deceived Adam and Eve in the garden of Paradise. While knowing fully well that they would lose their place with Allah and a peaceful home if they accepted his advice, so it will be with the so-called American Negroes, who after being offered heaven at once by Allah, accepts the wicked advice and the office of a temporary enjoyment with racial intermarrying and having sexual intercourse with a people with whom Allah is angry. (Foundation II, pg. 47)

The black man has ever suffered from his mistake in taking the white race for his friends. From that incident in Eden, six thousand years ago to this day, there is enmity and hatred between the two people (black and white). This cannot be removed except by the removal of one or the other race. The two could never come to and live up to a just agreement to do justice between each other. It is the impossible, unless their nature is changed. Adam and his wife accepted the advice of the devil against their Lord, and the serpent (the devil) has misled and put to death the prophets and the righteous servants of God ever since the fall of Adam. (Foundation II-76)

Ever since Adam (the Caucasian race) has been referred to as the man (the man made man or race and mankind); the world has taken it to include all men. *"No man hath seen God;" (I John 5:19)*. This is another confused teaching to the ignorant masses, for if no man hath seen God, then there is no God for you and me to look forward to seeing on the Judgement Day. If such is true, how will we know

Him since no one has ever seen Him? But we do know Him and what He looks like.

"We know that we are of God and the whole world lieth in wickedness." I John 5:19). (Foundation II, 152)

THE ADAM's (devil) children, the great trouble-makers, the demon, the fiend, Shaitan, the Adam's human beasts, the hell-raisers, the open arch-enemies of God, and all black mankind, who, in the beginning, disobeyed the law of God and introduced evil, filth and disrespect for God and His law of justice and righteousness among the nation of righteous (the black, brown, yellow and red people), are now on their traditional rampage against us, the so-called Negroes (their good old 400-year old slaves).

They have been murdering and raping us throughout the centuries; yet, you are foolish enough to love and adore them (the devils) after all of their evils poured upon you and me. It just does not make sense.

Do not be surprised at anything like evil that you see them do; only be surprise when you see them do an act of good in your favor. Evil is the nature of Adam's children. They even have you believing that you are from Adam which is absolutely false. Never say that you are from Adam. Adam was the father (devil) of sin and disobedience, the devil of you and me. (Foundation II - 171)

MURDERERS FROM THE BEGINNING

ACCORDING TO the word of Allah (God) and the history of the world, since the grafting of the Caucasian race six thousand years ago, they have caused more bloodshed than any people known to the black nation. They are born murderers made by nature to murder. The Bible and Holy Qur-an Sharrieff are full of teachings of this race of murderers. They shed the life blood of all the living, even their own. They are scientists at deceiving the black people. They deceive the very people of Paradise (Bible:Genesis 3:13). They kill their own brother (Genesis 4:8). The earth testified against them (Genesis 4:10) and revealed their guilt to its Maker. (Thy brother's blood crieth unto me from the ground). The very soil of America is soaked with the innocent blood of my people, the so-called Negroes, shed by this race of murderers, also crieth out to its maker for the burden of the innocent blood of the righteous slain upon her. Let us take a look at the devil's creation from the teachings of the Holy Qur'an.

"And when your Lord said to the angels, I am going to place in the earth One who shall rule, the angels said: What will thou place in it such as shall make mischief in it and shed blood, we celebrate Thy praise and extol Thy holiness." (Holy Qur-an Sharrieff 2:30).

This devil race has and still is doing just that - making mischief and shedding blood. As the Holy Qur-an says, "When your Lord said to the angels: Surely I am going to create a mortal of the essence of black mud fashioned in shape. (Holy Qur-an Sharrieff 15:28). The essence of black mud (the black nation) mentioned is only symbolic which actually means the sperm of the black nation, and they refuse to recognize the black nation as their equal, though they were made from them by one of their black scientists (names Yakub). They can never see their way in submitting to Allah (God). His religion, Islam and His prophets. (Foundation I - 159)

The white race (devils) was permitted to rule (Genesis 1:26) us not according to Justice and Righteousness, but by evil and falsehood. They were made wise enough by their father (Yakub) to deceive us in making falsehood to appear to be true, until the time and appearance of the Author of Truth (Allah), and Justice. They (the white race) make evil appear to be very attractive (Genesis 3:6) and in this way they capture many of the righteousness, for unlike attracts.

By nature they are murderers and liars (Genesis 4:8; John 8:44), therefore, they could not believe (John 8:37-40) in a true religion like Islam, which means man's entire Submission to the will of Allah (God).

Take a second look into the above said; if the devils were given the power to be the God of this world for a time (six thousand years), how could they by nature teach us to submit to Allah, and do the will of a God of Truth, freedom, justice and equality. Therefore, the people under a wicked ruler (the devil) of the world went astray from the True God and True Religion. They made for themselves all kinds of false gods and religions and bowed down and worshipped them. (Foundation I - 186)

The Bible itself, being tampered with and poisoned by them is made to read as though the salvation in it is for the white race. The so-called Negroes who are the real chosen people of God today are not mentioned under their name. The name "Israel" is used to blind the black people, especially the American so-called Negroes. For example: Mr. Herbert Armstrong of Pasadena, Calif., a radio and TV preachers, and Theodore Fitch of Council Bluff, Iowa., the author and writer of a small 56-page book titled: "Our Lord's Plan for the White Race," with a sub-title, "Who we are? Where we came from? Why he chose us? What our work is? Why our descendants must remain white?" These questions I will truthfully answer.

Mr. Fitch and others will agree that Adam and Jacob (Yakub) was the father of the white race. Adam, according to the Bible (Gen. 3:6) Holy Qur'an (2:30), was the first sinner and divine law-breaker, liar, mischief-maker, murderer and world deceiver. As for Jacob (Yakub) whose name was changed to Israel (Bible-Gen. 33:28), because he wrestled with the angel and did prevail, according to the world and teaching of Allah (God) to me, the only angel that Jacob (Yakub) wrestled with was the black man, to bring out of the black man the present white race - for the white race was and is in the germ of the black nation.

Mr. Yakub was the first to discover this germ in us and grafted this germ of the black man into an independent race of people. Mr. Yakub taught his people how to rule the black nation for 6,000 years. Their time expired in 1914, but they have been given an extension of time, until the black man could be awakened out of his 6,000 year sleep; so as to restore the American black people (the real lost-found members of their people) onto their own kind.

JACOB (YAKUB) the great deceiver who deceived his own father and brother (Gen. 27:19,35,36) robbed his father-in-law, Laban, of his cattle (Gen. 30:37-43, 31:12). (Foundation II - 214)

THE TRUTH that the world has been waiting to know is just the answers to the questions Mr. Fitch asked in his book titled, "Our Lord's Plan for the White Race" (the God of righteousness never had a plan for you), with a sub-title, "Who We Are?" (the enemies of God and the righteous); "Where We Came From?" (from the black man); "Why He Chose Us?" (the God of righteousness did not choose you; if He had, He would not destroy you); "What Our Work Is?" (evil); "Why Our Descendants Must Remain White?" (certainly, you should remain white).

The average so-called Negroes would be misled by Mr. Fitch and his answers to such questions if it were not that Allah (God) had visited us and revealed the true knowledge of the white race. The best of it all, He is making the white race manifest to the world of the original black nation that they are our enemies, the devils, and that their father (Yakub) that created them is also classified as the father

(Yakub) of the devil race, because of his idea and making of the devils. (Foundation II - 217)

YOU WILL KNOW THEM BY THEIR WORK

According to the word of Allah to me, the white race was created in haste. (See Chapter 21:37). Their father, Yakub, was in such a hurry to put his new people on earth as rulers, that he married them while very young (15 and 16 years of age). Knowing that they had only 6,000 years to go, he rushed them, so says the world of Allah to me. This makes them hasty by nature. They have a fiery temperament; while the original black man is meek and humble. They were made to attract the black people. With this attraction, and other actions of evil and filth, they intended to take you to hell with them. As they say: I will certainly lie in wait for them in Thy straight path; I will certainly come to them from before them and from behind them and from their right hand and from their left hand; and Thou shalt not find most of them thankful. And the devil swore to them both; most-surely, I am a sincere adviser to you." (7:12,16,17,21)

"Seek refuse from the evil of that which He (Yakub) has created." The father (Yakub) of this World created a World of evil, discord, and hate. If you do not agree with their evil doings, your goodness is then called hate or infidelity, and peace-breaker.

This world of Christianity has gone mad, and they think that every cry is against them. They are like robbers who have robbed, and are afraid that they will be recognized by their victim. Thieves know that light makes them Manifest. (Foundation II -98)

They don't tell you, but they know it, that when they were made - not created when they was made they was made - m-a-d-e, made and the Bible teaches you that they were made. Let us make man. But in the Holy Quran and Bible the make was condemned by the black scientist in those days. In the Quran, it's got it very good there, he questioned, the scientist questioned the maker of the white man in these words; "What will you make other than something to cause or rather to make mischief and cause bloodshed?" This he has done and is still doing. In the Bible, the first from Adam killed one of them before he was able to exercise any rules. (TOT 80)

This man being 6 years old when that he started his work. Now if he was 6 years old, he was on time, because he was going to make a people to live 6,000 years, and he did do just that as we see them and they call their number 6; and in the Bible they say their number is equal with the man. Well that's right, because we were ourselves created on the number 6 but it was not just a little 6 years, our went into billions and trillions, while his stopped here at just a few year carrying 12 months, 365 and 1/4 day. (TOT 71)

Brothers and sisters I think I've just about said enough and I think you have gotten to the birth of your worst enemy ever was on the earth, that's Yakub. Yakub the name, don't mean an enemy to us but the man that had that name he made an enemy for us, but you can be called Yakub or I can be called Yakub but that don't mean we are devils or no friend to the maker of that father of the devil Yakub. (TOT-75)

PRINCE OF DARKNESS & THE GOD OF THE WORLD

As we move on a little further into Yakub's work as a God, he was the God of this people and he made god's out of them as the Jesus teach you or tell you, that they are the God of this world - they are. Anything they want, they don't get on their knees to pray for it, they make it. They make it for you and me. Anything that the civilization needs, he produces it. This is what you call god-like power. He is god of this world. He can produce for the people their desire and material. (TOT-70)

The white race are the gods of their world. The white man puts us on wheels. You say, "Oh, but we had wheels before." That is all right if we did have wheels. He himself put you on a new wheel, one different from the one you were using for transportation. The white man moved you off the camel's back and the donkey's back and he put you onto fast-powered transportation such as automobiles and he put you on wings in the sky. All of this was new compared to what we had. So we would like to show you that the white man himself has made a new world. (OSHA PG119-120)

Allah (GOD) Who Came in the Person of Master Fard Muhammad, to Whom Praises are Due forever, taught me that every twenty-five thousand (25,000) years, each God Coming After the Other God made a new civilization. His

Belief, Teaching, and Theology were Different From the Other God Who Preceded Him Who Made a beautiful change in the History of the wisdom of man.

The making of the white race was something new. The difference between the two people (Black and white) is that the white race is a made people and they have a beginning and an ending. The Black Man is a created people and we have no beginning nor ending. (OSHA 119)

You use to think and was told that we all was created together alike, no we were not created together nor are we alike - they are different from us, and you are different from them. So don't get that idea that God created all from Adam. This is an ignorant teaching and misunderstanding that God made us all the same time and all are alike and God loves us all, you're mistaken. There were 2 gods in this kind of making: one made an evil man and the other one made a good man, or allowed a good man to remain. The evil man began 6,000 years ago and the good man, black people, have no birth record. God didn't create us 6,000 years ago, no sir. As God have taught me in the person of Master Fard Muhammad, to Whom praise is due forever, that we were here millions of years before white folks came, millions of years. There was no such thing as we all came together. Did you know that would have been a very tough job on the maker to make a black man and a white man at the same time? (TOT 332-333)

THE ORIGINAL OVERTAKES HIMSELF

Allah came to you and I today to operate on us and put the germ that is effecting us into the fire, the devil himself. You can not take the devil himself, today, and say you're going to exile him. He know every square inch of the earth almost, so you can't confine him in a place of exile. He doesn't only know the earth, he now knows the air. He would jumps up out of your cave now and go up in the sky. You can't put him in a cave unless you tie him down there so that he can't come out. If you're not careful, he may gnaw his way out. Anyway, he got away now. (SD Audio 71)

Often I am asked: (1) How can we live separately from the white people in the same country? (2) How can we be independent of them? (3) Would the white American agree to a separate territory for us in some suitable area where we could go for self and live alone in peace? (4) Would we get along in peace with each other?

The first question is very easy to answer: The black and white people, according to history, have always lived separately from each other. They were originally separated by God, and the Law of God forbids their intermixing, or intermarrying. The earth was originally inhabited by the Black Man.

Later on as colors (brown, yellow, red) began among them by the original scientists, (says the word of Allah to me), they became families to themselves, and from families to tribes; the term used today is "races" or "nations." Most of the black remained in the East along and below the equator of the earth.

The white race, after their creation 6,000 years ago, was given the part of earth today that is called Europe.

It is this race of people who actually started the sin of intermixing the two strange bloods, which is forbidden by God; just as they have disregarded all Divine Laws of Allah (God. They even seem to change the very natural laws of man in which he is made.

They first mixed with the original Black Arabs, then, as they traveled over the earth, for the past near 500 years, they mixed with brown, yellow and red races. They captured the blacks and made slaves of them. For the past 400 years, they have used them (so-called Negroes) for experimental purposes and even uses them now as one uses his tool! (Foundation I-166)

In the Bible you will find this, they say go down and take him out quick, lest he put forth his hands near the tree of life and he'll live forever. You never knew what this meant, because no one would teach you who is the tree of life. Do you think he would tell you - the black folks? No. The God that he was sending to black people was wrong, and to put forth his hands to the tree of life is only to go and intermix with us and get babies by them, that's laying hold to the tree of life. The tree of life is the black people - but he was wise. The book teach you that he knows good and evil. He was the tree the knows good and evil, which stood in the midst of the garden that had knowledge of the good and the evil, but was not doing good. He is the one that you'd better be careful. Don't you make a mistake, because he has a knowledge of you and himself too.

So, today God has decided to operate on the black man to get rid of the evil germ that's in him. The germ that will continually to come requires that we get rid of the white race.

I say let us be happy. Let Allah operate on the enemy. Get him out of us, put him in the fire where he belong. We cannot convert him to Islam, because we yet won't have a true Muslim. We will have to take the man and accept him for his faith if he has faith in Islam or in Allah, His religion and accept it for that, but we won't accept him in our religion to go to work to graft him back into the black man. We won't accept him for that. We're not going to graft him back into us. He has given us enough hell for the last few thousand years by coming in contact with us. He has poisoned 90 percent of the black people on earth, so Allah taught me, in the person of Master Fard Muhammad. You think that you can go to Arabia, yet, you won't find truth where you're suppose to find truth in all the Arabs. That's right. You will find the devil over there with the Arabs and he's sitting in the seat of authority.

I say to you, the black man needs to be operated on; operated on to get the devil out of us. This is the only way that we can rid ourselves of it. If the Arabs want to keep it and teach

that old doctrine that Islam is for everyone; yeah, in a way it is. Islam is offered to every man, although every man that is going to accept it, doesn't belongs to it. The black man by nature is born a Muslim. He doesn't have to accept it. He was born that, but the white race was not born a Muslim. They were born a wicked race and their father made them wicked to try us; to put him over us as our ruler. You will find in some of Muhammad's literature that every child that's born is born a Muslim. What's wrong Mr. Muhammad? Every snake that is born is a snake, he's not something else, he's a snake and every child that is born is not a black child. If he didn't come from a righteous parent how can he be righteous? (SD-71)
We have to refer to these things according to the fundamental truths of them and the principles upon which they are fashioned. Israel, we know is the whole of the white race. When we say Israel, it's the whole of the white race. We do know that according to the study and the nature of this people, they are a grafted people; we know that, and your scientists admit the same.

We cannot over look the fact that we can take the graft and the original and keep them together and leave one as a slave to the other after knowledge, you can't, you just can't do that. That's against the very law of nature for such, and if we are asking for the freedom to move here, we're doing nothing more than what you have once did according to the history of your race. It is nothing more than right that we try; for this is the nature of people. If I can't get along with you, I'll leave you. That's the nature of people. To force me to get along with you against my will, will not produce peace among you and I; no, because deep within I am still, probably poles apart from your way, or, we say, your thinking or doing.

There is a lot of evil in the world and if you would like to see it removed, how are you going to remove evil if you're going to try and associate evil with good? You can't remove evil by associating it with good. If you do, then you really don't want it removed.

If you have a sick patient of good and evil that you want to cure, could you cure that person of the disease without going at the root cause of disease? How could you cure him? You'll have to get to the cause. Alright, then the only way you can cure evil, you have to get after the cause. This is the way you look at it. Let me present my own personal observation and candid things about this; for in the root I get my ideas. If you try to eliminate the cause of evil, you have to go at the root of the evil cause. You have to know it, and recognize it as true, and remove the cause as a man knows it. Here sitting in my flesh are certain diseases, that he can medicate maybe for many years, and never cure, but if he go after the root of that thing, he could cure it overnight.

So it is with the people of today. All over the Earth there is dissatisfaction, and the dissatisfaction and disagreement is due to evil, and the only way to cure it now, since it's become universal or has universally spread and affect every human being, we have to go and diagnose the nation and find the cause. After diagnosing the nation, get rid of the cause. We can't say medicate it, because the prophets acted as that medication, and they were turned down in what they were worth, so now you must go and get the root of it, dig up the root, and throw it out, and then you'll be rid of it. (HISTN)

THE COMING OF THE SON OF MAN

The Black Man or gods of the Black Man are infinitely wise. They are being aroused to their Wisdom today to rule the people again throughout the ages of time.

The Present God's (Master Fard Muhammad's) Wisdom is infinite. No Scientist can see an end to This Man's Wisdom coming in the future. That is why the Bible and the Holy Qur'an refer to Him as The Greatest and Wisest of Them All and say that He will set up a Kingdom (Civilization) that will live forever. (OSHA PG 96-99)

That's exactly why, I said that this Israel or the scholars of Israel have this in their prophecy of the coming of a last man, and the nature in which he must come. This man that must come and must go after that which is lost. He must be a prepared man in order to fit the job well. He must be prepared. He must be part of both peoples. (HISTN)

[THIS SUBJECT OF THE "PREPARED ONE" CAN BE TAKEN UP IN THE BOOK, THE TRUE HISTORY OF MASTER FARD MUHAMMAD].

FOOTNOTE LEGEND

FOI - Flag of Islam, The;

FOUNDATION I - Elijah Muhammad, Foundation Years, Book I

FOUNDATION I - Elijah Muhammad, Foundation Years, Book II

HISTN - History of The Nation of Islam, Secretarius Audio Comp.

MTTBM - The Message To The Blackman In America

OSHA - Our Saviour Has Arrived Book

SD71 - Saviour's Day 1971, Operate On The Nation Audio Tape

TOT - The Theology of Time Book, Secretarius Audio Compilation

LTR - Letter to the Ministers, March 11, 1965.

The above books and audios are available at Secretarius MEMPS.

Thank you for purchasing this book. We trust the reading was rewarding and enlightening.

We offer various titles and a comprehensive collection of Messenger Elijah Muhammad's works. These works include:

- **Standard Published Titles**
- **Unpublished & Diligently Transcribed Compilations**
- **Audio Cassettes**
- **Video Cassettes**
- **Audio CD's**
- **DVD's**
- **Rare Articles**
- **Year Books**
- **Annual Brochures**

You are welcomed to sample a listing of these items by simply requesting a FREE archive Catalog.

Our contact information is as follows: